SpringerBriefs in Earth Sciences

More information about this series at http://www.springer.com/series/8897

J.H.L. Voncken

The Rare Earth Elements

An Introduction

J.H.L. Voncken
Civil Engineering and Geosciences
Delft University of Technology
Delft
The Netherlands

ISSN 2191-5369 ISSN 2191-5377 (electronic)
SpringerBriefs in Earth Sciences
ISBN 978-3-319-26807-1 ISBN 978-3-319-26809-5 (eBook)
DOI 10.1007/978-3-319-26809-5

Library of Congress Control Number: 2015956134

Printed on acid-free paper

This Springer imprint is published by SpringerNature
The registered company is Springer International Publishing AG Switzerland

Preface

This book was written as a summary of my long-standing interest in rare earth elements, which had started already when I was a master student at Utrecht University in the 1980s. Having begun seriously to tackle the topic early in 2009, it appeared that, during the so-called rare earth crisis from 2009 to approximately 2013, I was considered a kind of 'sole expert' in the Netherlands on rare earth element resources. This led me to being asked to give numerous presentations.

As a result of these many presentations, I realised that there was little or no information on the rare earth elements for the well-educated non-expert. So, in June 2014, after having given the umpteenth presentation on rare earth elements for an audience of members of the Groningen Department of the Royal Dutch Chemical Society (KNCV), I started with the preparations for writing this book.

This book is thus intended for the well-educated but non-expert reader, from any natural science discipline. This also explains the numerous footnotes scattered throughout the book, which are intended to elucidate the used terminology, or to supply short background information.

For the opportunity to write this book, I first of all would like to thank Dr. Mike Buxton, Head of the Resource Engineering Section, Delft University of Technology, who earnestly suggested me to write this work and allowed me to write it as part of my duties at Delft University. He also mentioned the importance of the South African Pilanesberg and Steenkampskraal deposits and supplied me with several papers and reports from the literature on these and other deposits.

Furthermore, this book could never have been written without the help and support of the following persons, companies, and institutions, which are listed here in alphabetical order:

- *Diana J. Bauer, Ph.D.,* US Dept. Of Energy, Director of the Office of Energy Policy Analysis and Integration Office of Energy Policy and Systems Analysis, for permission to use two images.
- *BeNature—The Green Living Channel*, USA for permission to use an image.
- *European Commission*, Directorate General for Internal Market, Industry, Entrepreneurship and SMEs for permission to use an image.

- *Anouk Borst M.Sc.* of the Geological Survey of Denmark and Greenland (GEUS), for supplying me with hard-to-get literature about the Illímaussaq Alkaline Complex, and the Igaliko Complex, South-West Greenland, and the REE deposits associated with them.
- *Dr. Denis Bykov,* Delft University of Technology, Reactor Institute Delft, for explaining the radioactivity of the element promethium (Pm).
- *Prof. dr.ir. Jan-Dirk Jansen*, Delft University of Technology, for introducing me to the Springer Briefs and supplying me with the contact at Springer Nature.
- *Mama's Minerals, Inc.*, Albuquerque and Santa Fe, New Mexico, USA, for permission to use an image.
- *Prof. Dr. Leo Kriegsman* and *Mrs. Arike Gill, M.Sc.,* of the Naturalis Biodiversity Centre at Leiden, the Netherlands, for supplying photographic images of rare earth minerals and rocks, for allowing me to make some photographs myself, and of course for permission to use all these images.
- *Peter and René van der Krogt,* from Delft, for permission to use their photographic images of the Ytterby mine.
- *Petra van Steenbergen and Hermine Vloemans,* my editorial partners at Springer Nature.
- *Dr. Phil Vardon,* Delft University of Technology, Department of Geosciences and Engineering, Section of Geo-Engineering, and an unknown reviewer for proofreading the chapters and correcting my sometimes Dutch-influenced English.
- *Dr. Yongxiang Yang,* Delft University of Technology, Department of Materials Science and Engineering, for supplying me with some literature on recycling of the rare earth elements.

Delft
October 2015

Jack Voncken

Contents

Chapter 1
The Rare Earth Elements—A Special Group of Metals

Abstract This chapter explains what the rare earth elements are, where and when they were discovered, and by whom. The name of each element is explained (as the elemental names are rather exotic), the misleading name for these metals, which suggests that they are rare (which they are not), is clarified, and the fact that they are not earth metals is established. The alkaline earth metals constitute group IIA in the periodic system, consisting of Be, Mg, Ca, Sr, Ba, and Ra; the rare earth elements are transition metals (group IIIB).

1.1 Introduction

Over the last few years, more and more people have become aware of a rather special group of metals: the rare earth elements. Many people have never heard of them, or probably just know them as "the upper one of these two rows of elements beneath the main body of the Periodic System of the Elements". Indeed, these metals, with a surprisingly large amount of applications, were rather obscure to the general public until 2009.

In 2009, China, which had almost a monopoly on the production of these elements (97 % of the world production), changed its position towards the world-wide rare earth market. China introduced production quotas, export quotas and export taxes, enforced environmental legislation, and granted no new rare earth mining licenses (Geschneider 2011). This caused world-wide anxiety among manufacturers of high tech equipment, because many of today's common (mostly) high-tech applications are not feasible without rare earth elements. Notable examples are hard-disk drives, smart phones, flat-screen televisions and monitors, rechargeable batteries (household and automotive) and tiny earphones. Others are lasers, strong permanent magnets for electrical generators, glass-polishing powders, and energy-saving lamps. This period, starting in 2009 and known as the "Rare Earth Crisis," made many people around the world aware of this peculiar group of elements.

J.H.L. Voncken, *The Rare Earth Elements*, SpringerBriefs in Earth Sciences,
DOI 10.1007/978-3-319-26809-5_1

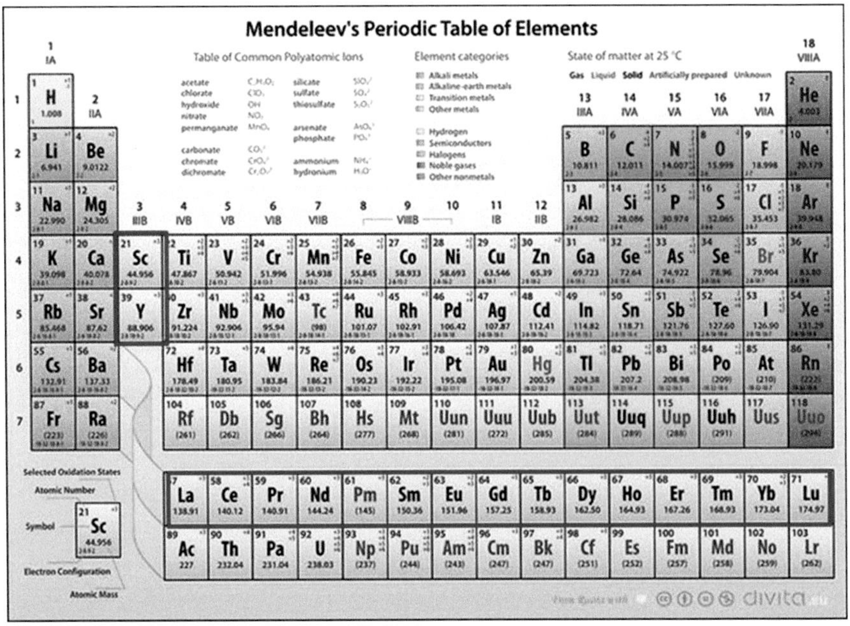

Fig. 1.1 The Periodic System of the Elements. *Reference* http://www.redbubble.com/people/seifip/works/5309681-mendeleevs-periodic-table-of-elements?p=poster. The REE including Sc and Y, are outlined in *red*

The rare earth elements are known under several names: rare earth metals, rare earths, or simply REE. They are a group of 17 strongly related heavy elements that comprise Sc, Y, and the Lanthanide Group. In Fig. 1.1, they are shown in the Periodic System of the Elements, outlined in red. The lanthanides should be positioned between the elements barium (Ba) and hafnium (Hf).

The discovery of the rare earth elements started at the end of the 18th century. The first element to be discovered was Yttrium, by Finnish chemist and mineralogist Johan Gadolin (Gadolin 1794, 1796; Weeks 1968; Gupta and Krishnamurthy 2005). See Fig. 1.2 for a portrait of Gadolin. By the end of the 19th century, all but two of the rare earth elements had been discovered. Lutetium was discovered in 1907, and the last one (promethium) only after the discovery of nuclear reactions. Promethium was identified in 1947 (Marinsky et al. 1947).

The rare earth elements are the elements *^{21}scandium (Sc), ^{39}yttrium (Y), ^{57}lanthanum (La), ^{58}cerium (Ce), ^{59}praseodymium (Pr), ^{60}neodymium (Nd), ^{61}promethium (Pm), ^{62}samarium (Sm), ^{63}europium (Eu), ^{64}gadolinium (Gd), ^{65}terbium (Tb), ^{66}dysprosium (Dy), ^{67}holmium (Ho), ^{68}erbium (Er), ^{69}thulium (Tm), ^{70}ytterbium (Yb), and ^{71}lutetium (Lu).*

The rather exotic names of these elements (compared to, for instance, lead and iron) will be explained later.

Fig. 1.2 Johann Gadolin. *Image Source* Wikipedia (2015) Johann Gadolin. *Reference* Dean and Dean (1996)

1.2 Atomic Structure

In the lanthanides (the elements La–Lu) the *f-orbitals*, which have 7 sub-orbitals, are filled. Each suborbital holds two electrons. As a result of this, there are 15 possibilities for filling the *f*-orbitals, giving rise to 15 lanthanide elements. These elements have closely related properties.

The lanthanides are all trivalent (3+), with the exception of cerium (which also forms 4+ ions), and europium, ytterbium, and samarium, which also will form 2+ ions.

The elements scandium and yttrium, which are also considered to belong to the rare earth elements (because of their similar chemical behaviour) also have a 3+ oxidation state. The atomic structure of the REE is further discussed in Chap. 3 (Physical and Chemical Properties of the Rare Earths).

1.3 Radioactivity

Of the rare earth elements, only one element is radioactive and has no stable isotopes. All other of these elements have stable isotopes, although for several elements, the isotopes have limited stability, but a very long half-life, so they may be considered stable. The radioactive element without a stable isotope is the element promethium (Pm). Promethium does not occur as a free element on Earth, whether as a metal or in compounds. It is synthetically manufactured in nuclear reactors, as it has practical applications. The radioactivity of promethium is further explained in Chap. 3.

1.4 Name

The term rare earth elements is a misnomer. They are ***not*** ***(Alkaline-)Earth Elements***: these are the elements in group 2 of the periodic system of the elements, which are the elements *beryllium* (Be), *magnesium* (Mg), *calcium* (Ca), *strontium* (Sr), *barium* (Ba) and *radium* (Ra). Instead, the rare earth Elements belong, as can be seen from Fig. 1.1, to the Transition Metals, group 3b. ***Neither*** are the rare earth elements ***rare*** (see below).

The name rare earth elements is closely associated with their discovery. Most of them were discovered in the 19th century, with the exception of yttrium (1794), lutetium (1907) and promethium (1943). Yttrium was discovered in 1794 by the Finnish mineralogist and chemist *Johan Gadolin*,[1] (shown in Fig. 1.2), in a mineral that was later named in his honor gadolinite.[2] Johan Gadolin is one of the very few scientists who have an element named in their honor. The element ***gadolinium*** was named after the mineral gadolinite, and thus the element gadolinium is indirectly named after Gadolin (Weeks 1968, pp. 684–685). Another naturally occurring element, which is indirectly named after a person, is ***samarium***. Samarium is named after the mineral samarskite, which itself is named after the Russian mining engineer *Vasili Samarsky-Bykhovets*, discoverer of the mineral.

But why are they called *rare*, and why *earth elements*? Well, in the 19th century, only one deposit of rare earth elements was known: a quarry near the town of *Ytterby*[3] in Sweden. Therefore they were thought to be rare. And *earth element*? Well, most REEs were first extracted as oxides, and in French (a major scientific language in the 19th century), an oxide of an element was known as the "terre" of that element, and "terre" literally also means "earth". Also in German, another major scientific language at that time, an oxide of an element was called the "Erde" (earth) of that element (see for instance Auer von Welsbach 1883).

As mentioned before, the rare earth elements are not rare either. Ore deposits of REE are quite restricted in numbers, but the abundance[4] of the elements is quite large. The most common rare earth element is cerium (Ce), which is, with a crustal abundance of 60 ppm, the 27th element in the Earth's crust, and has a larger abundance than, for instance, lead (Pb), the 37th element, which has a crustal abundance of 10 ppm. One of the least common rare earth elements (lutetium, crustal abundance

[1]Born: Åbo [now Turku], Finland, 5 June 1760; died: Wirmo, Finland, 15 August 1852.

[2]**Gadolinite** is a silicate with the formula $(Ce,La,Nd,Y)_2FeBe_2Si_2O_{10}$. Reference: Mindat.org.

[3]**Ytterby**: pronounce the "y" as the "e" in "to be", and the "e", as the "e" in "the". The last syllable should be stressed.

[4]**Abundance**: The abundance of a chemical element measures how relatively common (or rare) the element is, or how much of the element is present in a given environment by comparison to all other elements. **Crustal abundance of an element** is the estimate of the average concentration of that element in the continental crust.

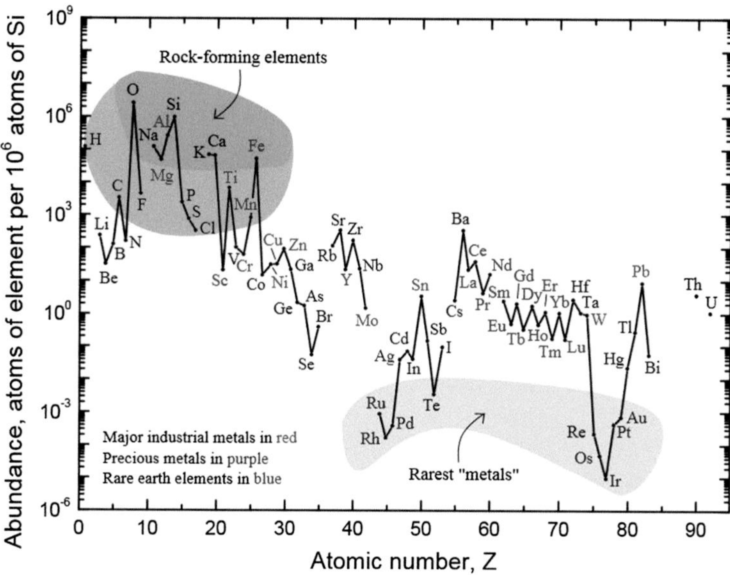

Fig. 1.3 Abundance of the elements given in atom fraction as a function of the atomic number. The rare earth elements are indicated in *blue*. Image from Haxel et al. (2005), courtesy of the U.S. Geological Survey

0.5 ppm), has a crustal abundance of about 200 times that of gold (0.0031 ppm) (Haxel et al. 2005).

In the REE series itself, a saw-tooth pattern can be seen (Fig. 1.3), which is due to the Oddo-Harkins-rule, that states that elements with an even atomic number have a higher abundance than elements with an odd atomic number (Oddo 1913; Harkins 1917)

1.5 Geochemical Behavior

Yttrium has geochemical behavior very similar to the lanthanides, which is why it is considered to be a REE. Scandium, however, shows geochemical behavior that is much more similar to that of the ferromagnesian transition elements (Fe, V, Cr, Co and Ni), due to its smaller atomic radius. This is also due to a different coordination in the crystal lattices of minerals. Therefore, scandium is often considered not to be a rare earth element, but a ferromagnesian trace element. In aqueous systems, however, scandium behaves more like the other REEs (McLennan 2012).

In magmatic systems, the REEs are in general lithophile elements ("rock-loving" elements). Also, they are "incompatible elements," which means that they tend to concentrate within the melt (magma), rather than in (early) crystallizing mineral phases. This is due to the large ionic radius of the REE (with the exception of scandium). As a result, yttrium and the lanthanides tend to be concentrated in late magmatic fluids and late crystallizing mineral phases (McLennan 2012).

1.6 The Deposit at Ytterby

The Ytterby mine was a feldspar mine (Swedish: *Ytterby feltspatgruva*), which originally was exploited for quartz. The quarry is situated in a granite pegmatite, and the mine was used as a source of feldspar and quartz for the porcelain trade with Great Britain and Poland. Quartz was quarried in this mine in the 1500s for iron-works in north Uppland (a historical province or *landskap* on the eastern coast of Sweden, just north of Stockholm). The quarrying of feldspar started at the end of the 1700s and continued until 1933, when the mine was shut down (van der Krogt 2014a, b). The location of the mine is shown in Figs. 1.4 and 1.5. Figure 1.6 gives a present day image of the mine, which is a historical landmark (Fig. 1.7).

The name *Ytterby* is composed from Swedish *ytter* = *outer*, and *by* = *village*. It thus literally means 'outer village'.

Most of the rare earth elements were discovered in the mineral *gadolinite*, (Ce,La, $Nd,Y)_2FeBe_2Si_2O_{10}$[5] (Mindat.Org) from Ytterby. The town has given name to yttrium, ytterbium, terbium, and erbium. Other elements discovered here are: gadolinium, holmium, thulium, scandium, lutetium, and tantalum (Source: Mindat.org).

The rare earth elements have rather exotic names. The origins of these names are explained in Table 1.1.

Below the discoveries of the rare earth elements are listed, and the year in which that happened (de Marignac 1878; Cleve 1879; Nilson 1879a, b, Gupta and Krishnamurthy 2005; Internet references are to the website of the *Royal Society of Chemistry, UK*).

- Sc L.F. Nilson & P.T. Cleve, 1879, (Cleve 1879; Nilson 1879b; http://www.rsc.org/periodic-table/element/21/scandium#history)
- Y J. Gadolin, 1794, C. Mosander, 1797, 1843, pure; S. West & B. Smith Hopkins, 1935; very pure. (Gadolin 1794, 1796; Weeks 1968) (http://www.rsc.org/periodic-table/element/39/yttrium#history)

[5]Gadolinite is currently differentiated in Gadolinite-(Ce) and Gadolinite-(Y), according to the most commonly occurring REE in the mineral. See also Mindat.org.

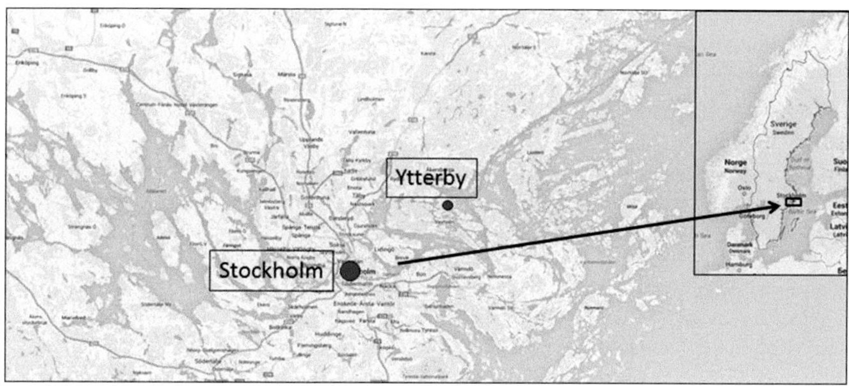

Fig. 1.4 The location of Ytterby. Extracted from Google Maps (2015)

Fig. 1.5 Ytterby on the island of Resarö. Extracted from Google Maps (2015)

- La C. Mosander, 1839; http://www.britannica.com/EBchecked/topic/330071/lanthanum-La?anchor=ref2246; (http://www.rsc.org/periodic-table/element/57/lanthanum#history; Weeks 1968)
- Ce H. Klaproth & J. Berzelius, 1803, W. Hisinger & C. Mosander, 1825, pure; (Weeks 1968) (http://www.rsc.org/periodic-table/element/58/cerium#history)

Fig. 1.6 The Ytterby mine in 2009. Photograph by van der Krogt (2014a, b)

- Pr and Nd C. Auer von Welsbach (1885). First they were considered to be one element, *didymium*, but as Delafontaine suspected, it was not one single element. It appeared to consist of two quite similar elements. The Greek word *didymos*, meaning *twin*, from which the name *didymium* is derived, refers to this. *Praseodymium* was isolated first, with "*Praseo-green*" referring to the green color of its salts. Then another one was isolated, which was the "*new one of the twin*": *Neodymium*[6] (http://www.rsc.org/periodic-table/element/59/praseodymium#history; Weeks 1968)
- Pm J.A. Marinsky, J.E. Glendenin, C.D. Coryell, 1947 (Marinsky et al. 1947)
- Sm P.E. Lecoq de Boisboudran, 1879 (Weeks 1968) (http://www.rsc.org/periodic-table/element/62/samarium)
- Eu E.A. Demarcay, 1886 (http://www.rsc.org/periodic-table/element/63/europium#history; Weeks 1968)

[6]The names were originally *Praseodidymium* and *Neodidymium.* The syllable *di* was later skipped (Gupta and Krishnamurthy 2005).

Fig. 1.7 The commemorative tablet placed in 1989 by the American Society of Metals. Photograph by van der Krogt (2014a, b)

- Gd J. Galissard de Marignac, P.E. Lecoq de Boisboudran, 1880, 1886 (http://www.rsc.org/periodic-table/element/64/gadolinium#history; Weeks 1968)
- Dy P.E. Lecoq de Boisboudran, 1886 (http://www.rsc.org/periodic-table/element/66/dysprosium#history)
- Ho M. Delafontaine and J.L. Soret, 1878 (http://www.rsc.org/periodic-table/element/67/holmium#history, Weeks 1968, Soret 1879).
- Tb C. Mosander, 1843 (http://www.rsc.org/periodic-table/element/65/terbium#history; (Weeks 1968)
- Er C. Mosander 1842 (impure), G. Urbain and C. James, 1905 (pure) (http://www.rsc.org/periodic-table/element/68/erbium#history; (Weeks 1968)
- Tm P.T. Cleve, 1879 (http://www.rsc.org/periodic-table/element/69/thulium#history; (Weeks 1968)
- Yb J. Galissard de Marignac, 1878 (de Marignac 1878; Nilson 1879a)
- Lu G. Urbain, 1907, independently: C. Auer von Welsbach, C. James, 1907 (http://www.rsc.org/periodic-table/element/71/lutetium#history; Weeks 1968)

Table 1.1 The etymology of the names of the rare earth elements

Atomic number	Symbol	Name	Etymology
21	Sc	Scandium	After the Latin word Scandia (Scandinavia), where the rare earth elements were first found
39	Y	Yttrium	After the village of Ytterby in Sweden
57	La	Lanthanum	From the Greek word "lanthanein", meaning hidden
58	Ce	Cerium	After the dwarf planet Ceres, itself named after the Roman goddess of agriculture and motherly love
59	Pr	Praseodymium	From the Greek "prasios", or green, and "didymos", meaning twin
60	Nd	Neodymium	From the Greek "neos", or new, and "didymos", meaning twin
61	Pm	Promethium	After the Greek god of fire Prometheus
62	Sm	Samarium	After the mineral samarskite, in turn named after Vasili Samarsky-Bykhovets (1803–1870), discoverer of samarskite
63	Eu	Europium	After the continent of Europe
64	Gd	Gadolinium	After the mineral gadolinite, in turn named after Johan Gadolin
65	Tb	Terbium	After the village of Ytterby in Sweden
66	Dy	Dysprosium	After the Greek "dysprositos", meaning "difficult to catch"
67	Ho	Holmium	After the medieval Latin name for Stockholm (Holmia)
68	Tm	Thulium	After the mythological, most northern place on Earth, Thule
69	Er	Erbium	After the village of Ytterby in Sweden
70	Yb	Ytterbium	After the village of Ytterby in Sweden
71	Lu	Lutetium	After Lutetia, the Latin name for Paris

Some details on the discoverers of the rare earth elements are listed in Table 1.2.

In the second half of the 19th century, the most common method by which elements were discovered was spectroscopy (Auer von Welsbach 1885, Gupta and Krishnamurthy 2005), after a long and extremely tedious separation procedure (described in detail in Auer von Welsbach 1883). Gadolin, for instance, did not have spectroscopy at his disposal, and actually did not separate the element yttrium in its pure form. He merely recognized the oxide of yttrium (yttria) as a new compound, but considered it to be an element. In the 18th century and until the first decade of the 19th century, "Earths" were considered to be elements (Gupta and Krishnamurthy 2005). It was Sir Humphrey Davy (1778–1829) who first isolated metals from "Earths" in 1807, by means of electrolysis (Gupta and Krishnamurthy 2005).

The names of the elements erbium and terbium were confused in the 19th century. Delafontaine used spectral analysis to prove the existence of erbium and terbium, but in his publications (probably unintentionally), he interchanged the

Table 1.2 The discoverers of the rare earth elements

	Name	Lifetime	(Co)-discoverer of	Country
1	Carl Auer von Welsbach, also mentioned as: Carl Auer, Baron von Welsbach	1858–1929	Praseodymium, 1885, neodymium, 1885, lutetium, 1907	Austria
2	Jöns Jakob Berzelius	1779–1848	Cerium, 1803	Sweden
3	Per Theodor Cleve	1840–1905	Thulium, 1879	Sweden
4	Charles DuBois Coryell	1912–1971	Promethium, 1947	USA
4	Marc Delafontaine	1837–1911	Holmium, 1878	Switzerland
5	Eugene Anatole Demarcay	1852–1903	Europium, 1901	France
6	Johan Gadolin	1760–1852	Yttrium, 1794	Finland
7	Jean Charles Galissard de Marignac	1817–1894	Gadolinium, 1880, ytterbium, 1878	Switzerland
8	Lawrence Elgin Glendenin	1918–2008	Promethium, 1947	USA
9	Charles James	1880–1928	Erbium (pure), lutetium, 1907	UK/USA
10	Martin Heinrich Klaproth	1743–1817	Cerium, 1803	Germany
11	Paul Emile Lecoq de Boisboudran	1838–1912	Samarium, 1879, dysprosium, 1886, gadolinium, 1886	France
12	Jacob Akiba Marinsky	1918–2005	Promethium, 1947	USA
13	Carl Gustav Mosander	1797–1858	Terbium, 1843, yttrium (very pure), lanthanum, 1839, erbium, 1843	Sweden
14	Lars Fredrik Nilson	1840–1899	Scandium, 1879	Sweden
15	B. Smith Hopkins	1873–1952	Yttrium (very pure)	USA
16	Jacques Louis Soret	1827–1890	Holmium, 1878	Switzerland
17	George Urbain	1872–1938	Erbium (pure), 1905, lutetium, 1907	France

names erbium and terbium as given by Mosander. The interchanged names have remained that way (Gupta and Krishnamurthy 2005).

In 1947, promethium, the last of the rare earth elements was discovered (Murphy 2006), and this closed a timespan of almost 150 years of laborious, painstaking chemical research to isolate and identify these fascinating elements (Szabadvary 1988).

References

Auer von Welsbach C (1883) Über die Erden des Gadolinits von Ytterby. Monatshefte für Chemie und verwandter Teile anderer Wissenschaften 4(1):630–642 (now Chem. Monthly)

Auer von Welsbach C (1885) Die Zerlegung des Didyms in seine Elemente. Monatshefte für Chemie und verwandter Teile anderer Wissenschaften 6(1):477–491

Cleve PT (1879) Sur Le Scandium. Comptes rendus hebd séances acad sci, Paris 89:419–422

de Marignac JG (1878) Sur l'ytterbine, nouvelle terre contenue dans la gadolinite. Comptes rendus hebd séances acad sci, Paris 78:578–581

Dean PB, Dean KI (1996) Sir Johan Gadolin of Turku: the grandfather of gadolinium. Acad Radiol 3(2):S165–S169

Forvo; http://www.forvo.com/word/ytterby. Pronunciation of the name Ytterby

Gadolin J (1794) Undersökning av en svart tong stenart ifran Ytterby stenbrott i Roslagen, Kungliga Svenska Vetenskapsakademien, Handlingar, pp 137–155

Gadolin J (1796) Von einer schwarzen, schweren Steinart aus Ytterby Steinbruch in Roslagen in Schweden, Crell's Annalen, (also: "Chemische Annalen für die Freunde der Naturlehre, Arzneygelährtheit, Haushaltungskunst und Manufacturen", or just "Chemischen Annalen"), pp 313–329. See for an online version: http://reader.digitale-sammlungen.de/de/fs1/object/display/bsb10072281_00321.html?contextType=ocr

Geschneider KA (2011) The rare earth crisis—the supply/demand situation for 2010–2015. Mater Matters 6(2):32–41

Google Maps (2015) https://maps.google.com/

Gupta CK, Krishnamurthy N (2005) Extractive metallurgy of the rare earths. CRC Press, Boca Raton, 484 pp

Harkins WD (1917) The evolution of the elements and the stability of complex atoms. I. A new periodic system which shows a relation between the abundance of the elements and the structure of the nuclei of atoms. J Am Chem Soc 39(5):856–879

Haxel GB, Boore S, Mayfield S (2005) U.S. geological survey. Fact Sheet 087-02. Rare earth elements—critical resources for high technology. http://pubs.usgs.gov/fs/2002/fs087-02/. Retrieved Oct 2014

Hisinger W (1838) Analyser af några svenska mineralier. 2. Basiskt Fluor-Cerium från Bastnäs. Kongl. Vetenskaps-Akademiens Förhandlingar 187-1891 (as Basiskfluor-cerium). (In Swedish)

Human Touch of Chemistry. http://www.humantouchofchemistry.com/famous.php?action=view&nid=859. Accessed July 2014

Marinsky JA, Glendenin LE, Coryell CD (1947) The chemical identification of radioisotopes of Neodymium and of element 61. J Am Chem Soc 69(11):2781–2785

McLennan SM (2012) Geology, geochemistry, and natural abundances of the rare earth elements. In: Atwood DA (ed) The rare earth elements—fundamentals and applications. Wiley, New York, pp 1–19

Mindat.org (http://mindat.org). Several references from this website: http://www.mindat.org/loc-3191.html, Gadolinite-Ce: http://www.mindat.org/min-1627.html, Gadolinite-Y: http://www.mindat.org/min-1628.html

Murphy CJ (2006) Charles James, B. Smith Hopkins, and the Tangled Web of Element 61. Bull Hist Chem 31(1):9–18

Nilson LF (1879a) Sur l'ytterbine, terre nouvelle de M. Marignac. Comptes rendus hebd séances acad sci, Paris 88:642–647

Nilson LF (1879b) Über Scandium, ein neues Erdmetall. Ber Dtsch Chem Ges 12(1):554–557

Oddo G (1913) Die Molekularstruktur der radioaktiven Atome. Journ Chim Phys 260–268

Royal Society of Chemistry. http://www.rsc.org. Data Retrieved Aug 2014

Soret JL (1879) Sur le spectre des terres faisant partie du groupe de l'yttria. Comptes rendus hebd séances acad sci, Paris 90(11):521–523

Szabadvary F (1988) The history of the discovery and separation of the rare earths. In: Gschneider Jr KA, Eyring L (eds) Handbook on the physics and chemistry of the rare earths, vol 11. Elsevier, Amsterdam, pp 33–80

van der Krogt PCJ (2014a) The discovery and naming of the rare earths, at http://elements.vanderkrogt.net/rareearths.php. In: Elementymology and elements multidict. http://elements.vanderkrogt.net/index.php. Accessed July 2014

van der Krogt PCJ (2014b) Photographs of the Ytterby quarry. Courtesy Peter and René van der Krogt, Delft. See also http://www.vanderkrogt.net/elements/indexes.php

Weeks ME (1968) Discovery of the Elements (7th edition, Ch. 16). J Chem Educ Am Chem Soc 667–699

Wikipedia (2014) Abundance of elements in Earth's crust. http://en.wikipedia.org/wiki/Abundance_of_elements_in_Earth%27s_crust. Accessed Oct 2014. See also U.S. Geological Survey (2005)

Wikipedia (2015) Johann Gadolin

Chapter 2
The Ore Minerals and Major Ore Deposits of the Rare Earths

Abstract This chapter gives an overview of the major and minor ore minerals of the rare earths, and of the related major ore deposits. As most of the rare earths are mined in China, the impression may arise that ore deposits of these metals occur in few other places on Earth. However, nothing is less true. The extensive overview of the ore deposits of the rare earths in this chapter is especially meant to indicate that deposits occur in quite a variety of countries, and that the apparent dominance of China is economically (and politically) powered.

2.1 Major Ore Minerals

At this date, the principal ore minerals for the REEs are *monazite*, *bastnaesite*,[1] and *xenotime*. The first REE mineral to be used was gadolinite, and from this mineral, several of the REE were first isolated, but it was not applied on an industrial scale. The first REE ore mineral from which REE were extracted for industrial use was monazite.

2.1.1 Monazite

Monazite (Breithaupt 1829) has a generalized chemical formula $CePO_4$. The name is derived from the Greek *monazeis*, meaning "*to be alone*" because of the isolated crystals of monazite, and the fact that it was quite rare when first found. For an image, see Fig. 2.1. Besides Ce, also other REE occur in monazite. These are mostly the LREEs (*L*ight *R*are *E*arth *E*lements: La, Ce, Pr, Nd, and Sm. Invariably, a mix of rare earths is present. The suffix Ce, La, Nd, or Pr is added to denote the

---

[1]Bastnaesite is also spelled bastnäsite, or bastnasite. In this book, the more common spelling bastnaesite will be used.

J.H.L. Voncken, *The Rare Earth Elements*, SpringerBriefs in Earth Sciences,
DOI 10.1007/978-3-319-26809-5_2

Fig. 2.1 Monazite, Iveland Setesdal, Norway. From the collection of Naturalis Biodiversity Center, Leiden, The Netherlands, Sample RGM412064. *Photograph* Naturalis. Used with permission

most frequently occurring REE. It usually also contains Th and/or U, but the amounts in monazite are generally too low to be extracted as a valuable by-product.

Monazite occurs generally as a minor mineral in granites and granodiorites and associated pegmatites, and also occurs in many metamorphic rocks.

Because monazite is:

(a) a heavy mineral with a specific gravity ranging between 4.8 and 5.5, with an average of 5.15 (Webmineral 2014), and
(b) very resistant to weathering,

it concentrates after weathering of the igneous or metamorphic host rock and subsequent transport in placers and heavy mineral sands (Gupta and Krishnamurthy 2005).

2.1.2 *Bastnaesite*

Bastnaesite (Fig. 2.2) was first described by the Swedish chemist Wilhelm Hisinger as "basis-fluor-cerium", from the Bästnas mine near Riddarhyttan, Västmanland, Sweden (Hisinger 1838). The general formula of bastnaesite is $Ce(CO_3)F$. Bastnaesite is another major REE ore mineral containing mostly the LREEs cerium, lanthanum, praseodymium, and neodymium. Of the HREEs, only Y is regularly found. A suffix Ce, La, Nd or Y is always added before the name to indicate the dominant REE. Low proportions of other HREEs are present. Also hydroxyl bearing versions exist: hydroxylbastnaesite-(Ce) and hydroxylbastnaesite-(Nd).

Bastnaesite, containing neither U nor Th, has replaced monazite as the primary LREE-ore mineral. Related minerals may arise from substitution of the fluorine and carbonate ions. Bastnaesite is a widespread mineral, although it never occurs in

Fig. 2.2 Bastnaesite (*yellowish material*), Mountain Pass California. Sample from the collection of Naturalis Biodiversity Center, Leiden, The Netherlands. Sample ST 82224. Photograph by J.H.L. Voncken. Used with permission

large quantities. It occurs in a variety of igneous rocks, such as carbonatites,[2] vein deposits, contact metamorphic rocks, and pegmatites (Gupta and Krishnamurthy 2005). Major ore deposits are generally related to carbonatite intrusions. Carbonatites are often found in relation to alkaline intrusives.

2.1.3 Xenotime

Xenotime (Fig. 2.3) was first described by Berzelius in a specimen from Hidra (Hitterø), Flekkefjord, Vest-Agder, Norway (Berzelius 1824, 1825). The name is derived from the Greek *xenos*—"*foreign*" and *time*—"*honor*". The generalized chemical formula of xenotime is (YPO_4). Xenotime, in contrast to monazite and bastnaesite, generally contains, besides Y, appreciable amounts of the HREE (*H*eavy *R*are *E*arth *E*lements: Y, Tb, Dy, Ho, Er, Tm, Yb, and Lu). Xenotime may contain up to 67 % REO, mostly the heavier elements (Gupta and Krishnamurty 2005). Most often occurring are dysprosium, ytterbium, erbium, and gadolinium. Xenotime contains lesser quantities of terbium, holmium, thulium, and lutetium. Xenotime is the major source for HREE (Table 2.1), but like monazite also contains Th and/or U, which, depending on the location of the deposit, and the concentration of these two elements in the mineral, may be a by-product or a pest.

For instance, Förster (1998a, b) list compositional variations of xenotimes and monazites from the German Erzgebirge/Fichtelgebirge, showing that xenotime

[2]Carbonatite is a rare igneous carbonate rock (almost invariably intrusive), consisting of more than 50 % carbonate minerals. Worldwide, only one example of extrusive rocks is known: the rocks and lavas of the (active) Ol Doinyo Lengai volcano, Tanzania.

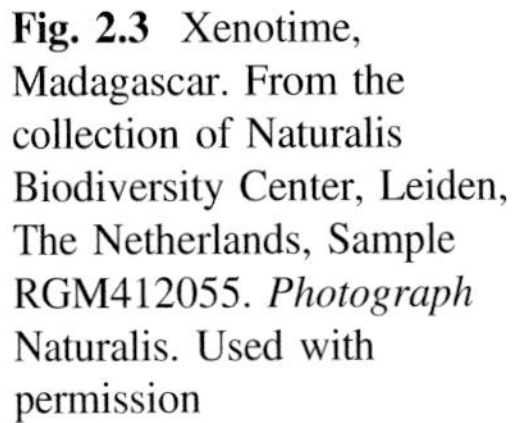
Fig. 2.3 Xenotime, Madagascar. From the collection of Naturalis Biodiversity Center, Leiden, The Netherlands, Sample RGM412055. *Photograph* Naturalis. Used with permission

tends to concentrate the HREEs (Gd, Tb, Dy, Ho, Er, Yb, Lu), whereas monazite tends to concentrate the LREEs (La, Ce, Pr, Nd, Sm).

Xenotime is an accessory mineral in pegmatites and other (non-basic) igneous rocks, but also is common in metamorphic rocks. Xenotime, being very similar to monazite (e.g. Gratz and Heinrich 1997), also has a high specific gravity, in the range 4.4–5.1, with an average of 4.75 (Webmineral 2014), and also concentrates in placers and heavy mineral sands. However, such deposits are not widespread (Gupta and Krishnamurthy 2005).

In Table 2.1 examples of typical compositions of monazite, xenotime and bastnaesite are given.

With respect to the actinides, monazite tends to concentrate thorium, whereas xenotime tends to concentrate uranium, but can take up also appreciable amounts of thorium. According to Deer et al. (2013), common varieties of monazite have 4–12 mol% of ThO_2, whereas uranium occurs in minor amounts. Van Emden et al. (1997) mention ThO_2 contents in monazite ranging 1.2–21.9 wt%, whereas UO_2 contents are from detection limit up to 0.75 wt%. Xenotime analyses show UO_2 contents ranging from detection limit to 5.82 wt%, while ThO_2 varies from detection limit to 8.44 wt%. Watt (1995) lists monazite compositions showing wt% of Th to vary from 5.17–21.41 wt%, and UO_2 from 0.22–3.17 wt%.

2.1.4 Eudialyte

Eudialyte (Fig. 2.4) was first described by Klaproth (1810). Eudialyte is a cyclosilicate with the general formula $Na_4(Ca, Ce)_2(Fe^{2+}, Mn^{2+})ZrSi_8O_{22}(OH, Cl)_2$. The name is from the Greek, meaning readily decomposable, referring to its easy dissolution in acids (Handbook of Mineralogy 2001; Anthony et al. 2014; Mindat.org 2014). Igneous eudialyte occurs in undersaturated alkaline intrusions and

Table 2.1 Examples of typical compositions of monazite, xenotime, and bastnaesite (Webmineral 2014)

Element	Monazite-Ce	Monazite-La	Xenotime-Y
La_2O_3	16.95	33.95	–
Ce_2O_3	34.16	17.10	–
ThO_2	5.50	5.50	–
P_2O_5	29.55	29.58	38.60
Nd_2O_3	14.01	14.03	–
Y_2O_3	–	–	61.40
CO_2	–	–	–
F	–	–	–
$O = F_2$	–	–	–
Total	100.17	100.17	100.00
Element	Bastnaesite-Ce	Bastnaesite-La	Bastnaesite-Y
La_2O_3	–	74.76	
Ce_2O_3	74.90	–	
ThO_2	–	–	
P_2O_5	–	–	
Nd_2O_4	–	–	
Y_2O_3	–	–	67.24
CO_2	20.08	20.20	26.21
F	8.67	8.72	11.31
$O = F_2$	−3.65	−3.67	−4.76
Total	100.00	100.00	100.00

Fig. 2.4 Eudialyte (*reddish material*), Kangerdlugssuaq fjord, Greenland. From the collection of Naturalis Biodiversity Center, Leiden The Netherlands, Sample RGM1055008. *Photograph* Naturalis. Used with permission

Table 2.2 Two typical eudialyte compositions (Handbook of Mineralogy 2001)

Element	Kipawa Lake, Canada wt%	Khibiny Massif, Russia wt%
SiO_2	50.35	50.14
TiO_2	0.38	0.46
ZrO_2	11.80	11.83
Al_2O_3	0.44	0.07
RE_2O_3	6.40	0.37
Fe_2O_3	0.19	0.50
Nb_2O_5	0.69	0.11
FeO	2.41	5.32
MnO	1.34	0.60
MgO	0.13	0.24
CaO	9.74	11.18
SrO	0.11	0.47
Na_2O	12.53	14.06
K_2O	0.43	1.39
F	0.23	
Cl	1.47	1.82
H_2O^+	1.64	1.07
H_2O^-		0.12
P_2O_5	0.03	
S		0.04
$-O = (F, Cl)_2$	0.43	0.41
Total	99.88	99.38

Table 2.3 REE-content of eudialyte (Harris et al. 1982)

REE_2O_3	wt%
Y_2O_3	3.61
La_2O_3	1.13
Ce_2O_3	2.37
Pr_2O_3	0.27
Nd_2O_3	1.12
Sm_2O_3	0.29
Gd_2O_3	0.69
Dy_2O_3	0.52
Er_2O_3	0.48
Yb_2O_3	0.18
Total	10.66

associated pegmatites (Deer et al. 1986a). Two typical eudialyte compositions are given in Table 2.2 (Handbook of Mineralogy 2001).

Harris et al. (1982) give the following concentrations for REEs in REE-rich eudialyte from Ascension Island (Table 2.3).

2.2 Minor REE Minerals

Other REE-minerals are numerous, but usually unimportant for industrial REE-extraction. In Table 2.4 a list of the other REE-containing minerals known up to now is given (Tasman Metals 2014a; Mindat.org 2014 Webmineral.com 2014). Like the major REE-minerals, for many minor REE-minerals several analogues

Table 2.4 Minor REE minerals

Mineral name	Formula
Aeschynite	(Ce, Ca, Fe)(Ti, Nb)$_2$(O, OH)$_6$)
Aenigmatite	(Na, Ca)$_4$(Fe, Ti, Mg)$_{12}$Si$_{12}$O$_{40}$
Allanite (Orthite)	(Ca, Ce)(Al, Fe^{2+})(Si_2O_7)(SiO_4)O(OH)
Ancylite	SrCe(CO_3)$_2$(OH)·(H_2O)
Apatite	$Ca_5(PO_4)_3F$ Apatite is as such not a rare earth mineral, but REEs may concentrate them, in which case they substitute for Ca
Brannerite	(U, Ca, Ce)(Ti, Fe)$_2$O$_6$
Britholite	Ca_2(Ce, Ca)$_3$(SiO_4, PO_4)$_3$(OH, F)
Cerite	(Ce, La, Ca)$_9$(Mg, Fe)(SiO_4)$_3$($HSiO_4$)$_4$(OH)$_3$
Cerianite	(Ce, Th)O_2
Cheralite	(Ca, Ce)(Th, Ce)(PO_4)$_2$
Churchite	YPO_4•2(H_2O)
Euxenite	(Y, Ce, Ca)(Nb, Ta, Ti)$_2$O$_6$
Fergusonite	Y(Nb, Ti)O_4)
Florencite	(Ce, La)Al_3(PO_4)$_2$(OH)$_6$
Gadolinite	$Y_2Fe^{2+}Be_2Si_2O_{10}$
Huanghoite	BaCe(CO_3)$_2$F
Hydroxylbastnaesite	(Ce, La, Nd)CO_3(F, OH)
Kainosite	Ce_2(Y, Ce)$_2$(Si_4O_{12})(CO_3)·H_2O
Loparite	(Na, Ce, Ca, Sr, Th)(Ti, Nb, Fe)O_3
Mosandrite	Na(Na, Ca)$_2$(Ca, Ce, Y)$_4$(Ti, Nb, Zr)(Si_2O_7)$_2$(O, F)$_2$F$_3$
Parisite	Ca(Ce, La)$_2$(CO_3)$_3$F$_2$
Rinkite	(Na, Ca)$_3$(Ca, Ce)$_4$Ti(Si_2O_7)$_2$OF$_3$
Samarskite	(Y, Fe^{3+}, U) (Nb, Ta)$_5$O$_4$
Synchisite	Ca(Ce, Nd, Y)CO_3F
Steenstrupine	$Na_{14}Ce_6Mn^{2+}Mn^{3+}Fe_2^{2+}(Zr, Th)(Si_6O_{18})_2(PO_4)_7 \cdot 3(H_2O)$
Tengerite	$Y_2(CO_3)_3$•2–3(H_2O)
Thalenite	$Y_3Si_3O_{10}$(OH)
Yttrotantalite	(Y, U, Fe)(Ta, Nb)O_4
Zircon	$ZrSiO_4$. Zircon also is not a rare earth mineral as such, but like apatite, it may concentrate REEs, in which case they substitute for Zr

exist, because several of the REEs may substitute for one another. These minerals are then indicated with a suffix for the predominating REE, and every single one is another mineral. For instance, for *aeschynite,* there are defined: *aeschynite-Ce, aeschynite-Nd,* and *aeschynite-Y.*

Allanite is a disilicate (sorosilicate) mineral from the epidote group, with the general formula (Ca, Ce)(Al, Fe^{2+})(Si_2O_7)(SiO_4)O(OH). It is named for the Scottish mineralogist Thomas Allan (1777–1833). Allanite was described by Thomson (1810). Allanite is also called orthite (synonym). Igneous allanite occurs in granite, granodiorite, monzonite and syenite intrusions. Allanite may contain up to 20 wt% *of REE,* expressed as REE_2O_3. This appears to be mostly $La_2O_3 + Ce_2O_3$, although from the LREEs also Nd_2O_3 is reported in concentrations up to approximately 3 wt % (Deer et al. 1986b). Other REEs occurring in allanite are: Dy_2O_3 (up to 0.21 %, although Deer et al. also report one analysis from 1940 with an unlikely amount of 6.96 wt%), Sm, and Gd (up to 1.2 wt% REE_2O_3). The REEs replace Ca. Allanite may contain up 5 wt% ThO_2. UO_2 occurs up to 0.5 wt% (Deer et al. 1986b, 2013).

2.3 REE-Containing Rocks

REE are often associated with a carbonate rock called carbonatite: an igneous (generally intrusive) rock consisting of more than 50 % carbonate minerals (carbonates of Ca, Mg, Na and Fe). Carbonatites as such are closely related to continental alkaline rocks (McBirney 1993). These are a group of rare (alkaline) igneous rocks that have an ultramafic or mafic silica-depleted character. Alkaline rocks are important for their high concentrations of incompatible or so-called Large-Ion-Lithophile elements (LIL-elements). Among others, these elements are the metals niobium (Nb), tantalum (Ta), and the rare earth elements (REE). Alkaline igneous rocks occur in several geological settings.

2.4 Ore Deposits of the Rare Earths

The deposits mentioned here are currently (2015) the most well-known economically interesting REE deposits. An extended list of economic and non-economic REE-occurrences can be found in Orris and Grauch (2002). Recently, ion adsorption deposits, quite similar to those in southern China have been discovered on the island of Madagascar (Desharnais et al. 2014). Also it is likely that more deposits will be found in the future. Deposits described in this work are listed in Table 2.5. The descriptions given here are of course by no means exhaustive.

Table 2.5 REE-deposits described/mentioned in this chapter

Deposit	Location	Type	Main REE	REE-mineral(s)
Mountain Pass	California, USA	Carbonatite	La, Ce, Nd	Bastnaesite
Bayan Obo	Inner Mongolia, China	Carbonatite/hydrothermal	LREE	Bastnaesite, parisite, monazite
Mount Weld	SW-Australia	Laterite/carbonatite	LREE	Apatite, monazite, synchysite, churchite, plumbogummite-group minerals
Ilímaussaq (*Kvanefleld*, *Kringlerne*, *Motzfeldt Sø*)	Greenland (Denmark)	Peralkaline igneous	La, Ce, Nd, HREE	Eudialyte, steenstrupine
Pilanesberg	South Africa	Peralkaline igneous	Ce, La	Eudialyte
Steenkampskraal	South Africa	Vein	La, Ce, Nd	Monazite, apatite
Hoidas Lake	Canada	Vein	La, Ce, Pr, Nd	Apatite, allanite
Thor lake	Canada	Alkaline igneous	La, Ce, Pr, Nd, HREE	Bastnaesite
Strange Lake and Misery Lake	Canada	Alkaline igneous / hydrothermal	La, Ce, Nd, HREE	Gadolinite, bastnaesite
Nolans Bore	Australia	Vein	La, Ce, Nd	Apatite, allanite
Norra Kärr	Sweden	Peralkaline igneous	La, Ce, Nd, HREE	Eudialyte
Khibina and Lovozero	Russia, Kola Peninsula	Peralkaline igneous	LREE + Y, minor HREE	Eudialyte, apatite
Nkwombwa Hill	Zambia	Carbonatite	LREE	Monazite, bastnaesite
Kagankunde	Malawi	Carbonatite	LREE	Monazite-Ce, bastnaesite-Ce
Tundulu	Malawi	Carbonatite	LREE	Synchesite, parisite, bastnaesite
Songwe	Malawi	Carbonatite	LREE, especially Nd	Synchysite, apatite

(continued)

Table 2.5 (continued)

Deposit	Location	Type	Main REE	REE-mineral(s)
Chinese ion adsorption deposits	South China	Soils	La, Nd, HREE	Clay minerals
Maoniuping	Sichuan, China	Carbonatite	LREE	Bastnaesite
Dong Pao	Vietnam	Carbonatite	LREE	Bastnaesite, parisite
Deep Sea	Pacific Ocean	Crusts, pelagic muds		

Fig. 2.5 The location of the Mountain Pass deposit (*orange balloon*). Modified after Google Maps (2015). The city of Los Angeles in the center on the bottom of the map

2.4.1 Mountain Pass

For a long time, the major ore deposit for rare earths was the Mountain Pass deposit, in California, USA (Fig. 2.5).

Geological prospecting in the area started already in 1861, but the most important discovery, that of the rich REE-deposit, was made much later. During a prospect for uranium in 1949, a strongly radioactive outcrop of a vein, close to the shaft of the Sulphide Queen lead and gold mine, was found. The vein contained a large amount of a light-brown heavy mineral, identified as bastnaesite (Fig. 2.2). Finding many other veins in the area, and all of the material being radioactive, it was concluded that a large amount of bastnaesite, associated with some thorium-bearing minerals, must be present in the area (Hewett 1954).

The Molybdenum Corporation of America[3] bought the rights to the ore bodies in February 1950. Later that year, the company started to sink a shaft (Hewett 1954). Many more veins with rare earth minerals were found in the surrounding hills.

In 1950, J. C. Olson of the USGS and D.F. Hewett started to make a geological map of the area. They also found a large body of a carbonate type of rock, with relatively high amounts of barite ($BaSO_4$), about 20 %. The carbonate rock mass was called the Sulphide Queen carbonate body. Laboratory investigations showed

[3]The Molybdenum Corporation of America changed its name to *Molycorp* in 1974. The corporation was acquired by *Union Oil* in 1977, which in turn became part of *Chevron Corporation* in 2005. (Molycorp 2014; www.molycorp.com). In 2008, Chevron sold the Mountain Pass mine to the privately held *Molycorp Minerals LLC.*

Table 2.6 Mean composition of Mountain Pass concentrate (Castor 2008)

Oxide	(wt%)
La_2O_3	33.79
CeO_2	49.59
Pr_6O_{11}	4.12
Nd_2O_3	11.16
Sm_2O_3	0.85
Eu_2O_3	0.105
Gd_2O_3	0.21
Tb_4O_7	0.016
Dy_2O_3	0.034
Ho_2O_3	0.034
Er_2O_3	0.006
Tm_2O_3	0.002
Yb_2O_3	0.002
Y_2O_3	0.13
Total	100.049

Fig. 2.6 Sövite. Alnö alkaline and carbonatite ring complex, Sweden. From the collection of Naturalis Biodiversity Center Leiden, The Netherlands, Sample RGM387008.0. *Photograph* J.H. L. Voncken. Used with permission

that a large proportion of the entire barite-carbonate rocks contains 5–15 % bastnaesite.

Although for the carbonate rocks at Mountain Pass alongside a magmatic origin, a sedimentary origin was considered, it finally became clear that the carbonate body was of igneous origin (Olson et al. 1954).

The mean composition of concentrate from Mountain Pass is given in Table 2.6. Figure 2.6 shows an example of a carbonatite rock type, called sövite.

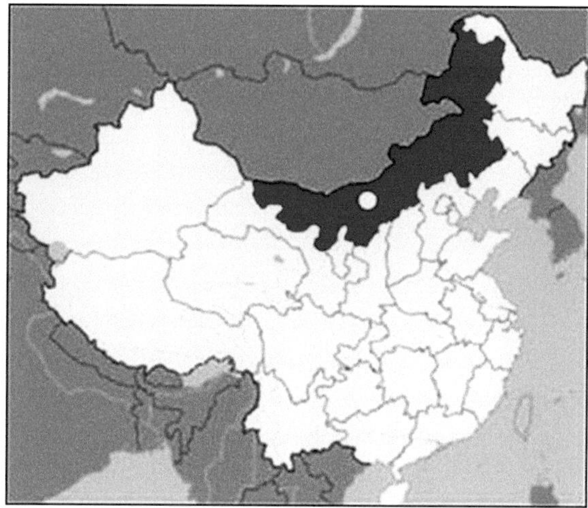

Fig. 2.7 Location of the Bayan Obo deposit. Modified after Drew et al. (1990) and Wikipedia (2014)

2.4.2 *Bayan Obo, Inner Mongolia, China*

Inner Mongolia is an autonomous region of China, bordering Mongolia and Russia. The location of Bayan Obo is indicated in Fig. 2.7. It is the largest mining town in the province. The name Bayan Obo is also spelled Baiyun-Obo or Baiyun Ebo (Encyclopedia Brittanica online 2014; Mindat.org 2014).

The deposit is located in the sediments of the Bayan Obo group. This group of rocks consists of low-grade metamorphic sandstones, siltstones, limestones, and dolomites. The ore deposits occur in a synclinal structure, termed the Bayan syncline (or Kuan syncline, Yang et al. 2009), in meta-sedimentary dolomite[4] marble with quartzite beds (in local stratigraphy called the H8-unit), which in turn is covered by shale. This dolomite marble is of Middle Proterozoic age (Chao et al. 1997). The ore bodies at Bayan Obo occur as hydrothermal replacements of the dolomite marble. The shale overlying the dolomite apparently acted as a seal for the later hydrothermal fluids that caused the mineralization (Drew et al. 1990).

It must also be mentioned here that the Bayan Obo ores are not only REE-ores. There also is significant mineralization of iron ore (Fe) and of niobium (Nb). Ore reserves are: approximately 1500 million metric tons of iron ore with an average grade of 35 %; 1 million tons of niobium ore with an average grade of 0.13 %; and at least 48 million tons of rare earth oxides with an average grade of 6 % (Drew et al. 1990). The deposit was discovered in 1927. Mining, not for REE or Nb, but for iron began in 1957.

[4]Dolomite rock is defined as a carbonate rock, containing calcite <50 %, dolomite >50 %. (Visser 1980). The term dolostone has been coined to avoid confusion with the mineral dolomite, $CaMg(CO_3)_2$, but has not gained general acceptance.

Table 2.7 Average REE-composition of Bayan Obo ore (Zhi Li and Yang 2014)

Oxide	wt%
La_2O_3	24–26
CeO_2	50
Pr_6O_{11}	3–5
Nd_2O_3	16–18
Sm_2O_3	1.5
Eu_2O_3	0.2
Gd_2O_3	0.4
$(Tb–Lu)_2O_3$	0.2–0.3
Y_2O_3	0.3

Most researchers agree that the ores themselves are formed by hydrothermal alteration (Yang et al. 2009; Yang and Le Bas 2004; Drew et al. 1990; Qiao et al. 1997; Chao et al. 1997). The origin of the hydrothermal fluid is either a mantle derived carbonatitic magma (Zhang et al. 2002) or related to (carbonatitic) magma generation as a result of subduction (Drew et al. 1990).

Wang et al. (1994) found an age window of 450–555 Ma for the formation of the deposit, which is approximately early Cambrian to late Ordovicium.

The average composition of the ore is given in Table 2.7.

2.4.3 Mount Weld, South-West Australia

Mount Weld is located 35 km south of Laverton, Western Australia (Fig. 2.8). The rare-earth-element deposit was discovered in 1988, but was not exploited for a long time, because China was supplying REE at low prices. When the so-called REE-crisis (see Chap. 6) came about in 2009, Lynas Corporation,[5] the company owning the deposit, decided to exploit the rich REE-deposit.

In January 2012, the Mineral Resource estimate for Mount Weld was 23.9 million tons, at an average grade of 7.9 % REO, for a total of 1.9 million tons REO.

The Mount Weld carbonatite dates from the Proterozoic (approximately 2025 Ma ago) and is emplaced in the Yilgarn Craton (Hoatson et al. 2011). The carbonatite is covered by a thick lateritic soil, which in its turn is covered by lacustrine and alluvial sediments (Lottermoser 1990). The Mount Weld deposit occurs in the weathered layer.

The primary igneous REE-bearing minerals are apatite, monazite, and synchysite (Willett et al. 1989; Lottermoser 1990). Secondary minerals containing REEs are

[5]https://www.lynascorp.com/Pages/home.aspx.

Fig. 2.8 The location of Laverton in SW-Australia. Modified after Google Maps (2015)

Table 2.8 Average composition of the ore from Mount Weld (Lynas Corporation 2013)

Oxide	wt%
La_2O_3	25.50
CeO_2	46.74
Nd_2O_3	18.50
Pr_6O_{11}	5.32
Sm_2O_3	2.27
Dy_2O_3	0.12
Eu_2O_3	0.44
Tb_2O_3	0.07

Fe-oxyhydroxides, secondary monazite, churchite, and plumbogummite[6]-group minerals.

At Mount Weld, the bulk of the REE and Y are incorporated in secondary monazite, churchite and a plumbogummite-group mineral. The LREE preferentially enter (secondary) monazite, whereas the HREEs are mostly found in churchite.

The average composition of the ore is given in Table 2.8.

[6]Plumbogummite is $PbAl_3(PO_4)_2(OH)_5 \cdot H_2O$.

2.4.4 Ilímaussaq Alkaline Complex, South Greenland

Of a completely different nature is the Ilímaussaq complex in South-West Greenland. The complex is related to so-called alkaline igneous rocks. Alkaline igneous rocks are defined in terms of their alkali ($Na_2O + K_2O$) content and silica content. For definitions see for instance Sörensen (1974) and Streckeisen (1967, 1980). In terms of volume, alkaline igneous rocks make up less than one percent of all igneous rocks (Fitton and Upton 1987). They have, however, been studied intensively, because of their peculiar mineralogy, and the association of these rocks with so-called *large ion lithophile elements* or LILE's. Among these elements are *niobium, tantalum,* and the *rare earth elements.* Also they may be associated with economic deposits of *apatite* and *diamonds* (Fitton and Upton 1987). The most important occurrences of alkaline igneous rocks are found in continental rift valleys. The best studied occurrence is the Proterozoic Gardar Province in West Greenland, containing the Ilímaussaq complex.

The Ilímaussaq Complex is the type locality of the so-called *agpaitic magmatism*. Agpaitic rocks are peralkaline nepheline syenites with complex Zr, Ti and REE minerals, such as eudialyte, steenstrupine, rinkite and mosandrite, rather than for instance zircon and ilmenite (Sörensen 1992, 1997). Other prominent minerals are aenigmatite[7] and astrophyllite.[8]

The age of the complex has been determined by several researchers. Ages found range between: 1130 ± 50 Ma, and 1168 ± 21 Ma (Sörensen 2001; Blaxland et al. 1976). This latter age is on the border between the Paleoproterozoic and the Middle Proterozoic.

Ilímaussaq has large resources of uranium, rare earth elements and zinc. The complex is located close to existing infrastructure in southern Greenland, making exploitation of the mineral resources favorable (Greenland Minerals and Energy 2014). The location of the complex is shown in Fig. 2.9. In the Ilímaussaq complex, we find two very high-quality ore deposits for rare earth elements: Kvanefjeld and Kringlerne. The nearby Igaliko nepheline syenite complex (age: 1273 Ma) hosts the Motzfeld deposit.

2.4.4.1 Kvanefjeld

Kvanefjeld was discovered in the 1950s. and was extensively studies during the 1960s, 70s and 80s. The main REE-ore mineral is steenstrupine. The Kvanefjeld REE deposit is dominated by Ce (approx. 40 %), La (approx. 25 %), Nd (approx. 15 %), Y (approx. 10 %), Pr (approx. 5 %), with the HREE making up the remaining 5 %.

---

[7]Aenigmatite is $Na_2Fe_5^{2+}TiSi_6O_{20}$.

[8]Astrophyllite is $K_2Na(Fe, Mn)_7Ti_2Si_8O_{26}(OH)_4$.

Fig. 2.9 Greenland with the location of Ilímaussaq. Modified after Google Maps (2015)

The total amount of REE-resources is 619 Mt, composed of both indicated resources of 437 Mt and inferred resources of 182 Mt (Sørensen and Kalvig 2011).

2.4.4.2 Kringlerne (a.k.a. Tanbreez)

The Kringlerne deposit occurs in the lower cumulates of the layered agpaitic nepheline syenites. The cumulates consist of 29 cyclic layers, which amount to a total thickness of about 200 m. They are composed of black arfvedsonite dominated syenite, reddish eudialyte dominated syenite and whitish feldspar dominated syenite (Sørensen and Kalvig 2011).

The deposit contains 1000 Mt grading 2 % ZrO_2, 0.25 % Nb_2O_5, 0.5 % REO, 0.1 % Y_2O_3 and 0.025 % Ta_2O_5. The distribution of light and heavy REEs in eudialyte is reported to be 88 % and 12 % respectively (GEUS 2011). Kringlerne is considered to be very rich in HREE (Tanbreez 2014).

The deposit is now also referred to as Tanbreez, after the company that is exploiting it (Tanbreez Mining Greenland A/S). The name ***Tanbreez*** is an acronym, composed of ***Ta*** for tantalum, ***nb*** for niobium, ***ree*** for REE, and *z* for Zr, zirconium.

These are the metals that are intended to be mined from the deposit (Tanbreez 2014, 2015).

2.4.4.3 Motzfeldt REE Deposit

The Motzfeldt Centre within the Igaliko complex can be found some 60 km to the north-east of the Ilímaussaq complex. It has an age of 1273 Ma, and is rather similar to Ilímaussaq, in consisting of agpaitic rock sequences. A zone of pegmatites and hydrothermal alteration created a mineralized zone which was in the past intensively explored for assessing possible Nb-Ta-mineralisation. In 2012 re-evaluation started concerning its REE potential (Steenfelt 2012).

The rock units of the Motzfeld deposit are highly variable in texture and mineralogy, and contain very high concentrations of Th, U, Nb, Ta, Zr and REE and volatile components such as F and H_2O. The inferred roof zone is the most extreme in diversity of rock types and enrichment of the elements mentioned above (Finch et al. 2001; Tukiainen 2014).

2.4.5 Pilanesberg, South Africa

The Pilanesberg (also called Pilansberg) is an agpaitic ring intrusion along a contact between granitic and noritic units of the Bushveld Complex, South Africa. The Pilanesberg complex dates from the Mid-Proterozoic (approximately 1250 ± 50 Ma ago). The complex is made up of phonolitic to trachitic pyroclastic rocks and lavas. These are intruded by diverse nepheline syenites, tinguaite[9] dikes and large cone sheets. It occurs along the noritic and granitic phases of the Bushveld complex (Ribeiro Olivo and Williams-Jones 1999). The Pilanesberg is located some 120 km NW of Pretoria. Presently it is a National Park and Game Reserve. See Fig. 2.10.

One of the major rock types, called the “green syenite”, consists of almost 20 % eudialyte, which is the main REE-carrier in the rock (Ribeiro Olivo and Williams-Jones 1999).

2.4.6 Steenkampskraal, South Africa

The Steenkampskraal mine is located about 350 km north of Cape Town (see Fig. 2.11) in the Northern Cape province. The mine was first operated for thorium and also for REE by a subsidiary company of Angalo American from 1952 to 1963.

[9]Tinguaite is the dike equivalent of phonolite.

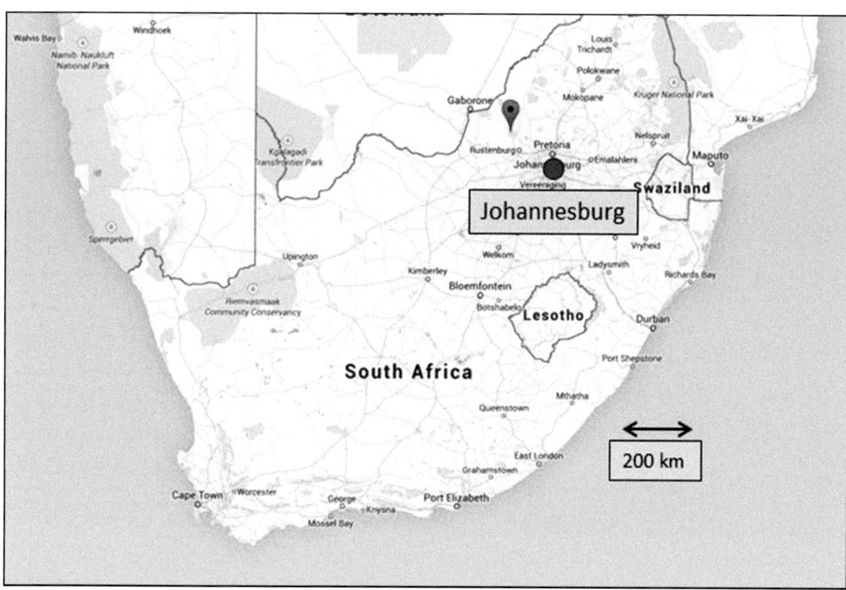

Fig. 2.10 The location of the Pilanesberg deposit, South Africa. Point of *orange balloon* is at the location of the Pilanesberg. Modified after Google Maps (2015)

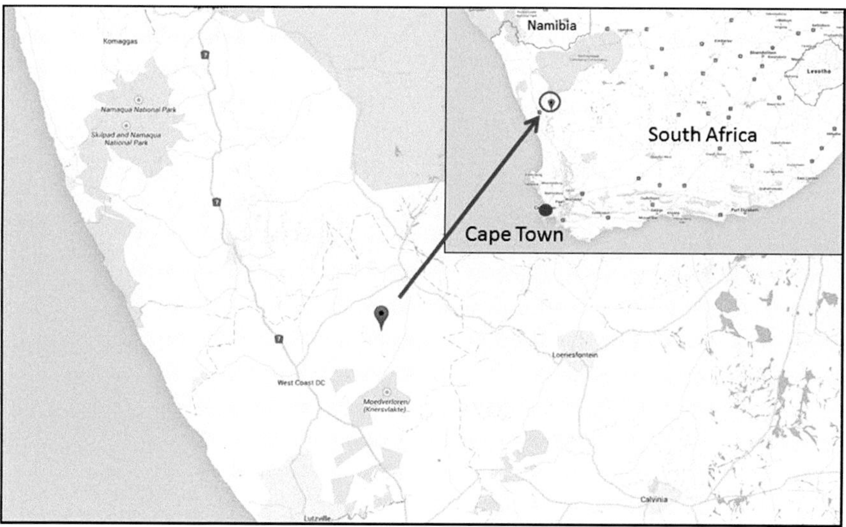

Fig. 2.11 The location of the Steenkampskraal deposit. Modified after Google Maps (2015)

Table 2.9 REE-composition of material from the mine

Element	wt%
Y_2O_3	0.91
La_2O_3	4.78
CeO_2	10.5
Pr_6O_{11}	1.18
Nd_2O_3	4.1
Sm_2O_3	0.643
Eu_2O_3	0.014
Gd_2O_3	0.436
Tb_4O_7	0.052
Dy_2O_3	0.221
Er_2O_3	0.065
Tm_2O_3	0.006
Yb_2O_3	0.025
Lu_2O_3	0.003
Total	22.935

The mine was closed in 1963, but has been re-opened because it provides an important REE source (Andreoli et al. 1994).

The current operator of the mine is Steenkampskraal Monazite Mine (Pty) Ltd. (SMM), which is 75 % owned by Rare Earth Extraction Company ("Rareco"). This company is fully owned by the Great Western Minerals Group.

The Steenkampskraal monazite district is located within the southern part of the Namaqua-Natal Metamorphic province. The age of this area is Middle Proterozoic. The largest town is Springbok.

Mineralization occurs in an ore vein, which is predominantly composed of monazite, with small amounts of allanite, REE-bearing xenotime, apatite and thorite. Mineralized material consists mainly of monazite enriched by cerium and lanthanum, but all rare earth elements are present, including yttrium. The average composition of the rocks is 45 wt% REE-oxides, 4 wt% ThO_2, 18 wt% P_2O_5, 1 wt % Cu, 0.1–1.5 wt% ZrO_2. U_3O_8 occurs up to 600 ppm. Traces of gold are also present.

The REE composition of material from the mine (Jones and Hancox 2012) is given in Table 2.9.

2.4.7 Hoidas Lake, Canada

The deposit is located in the southern part of the Rae province in northern Saskatchewan, Canada (Fig. 2.12). The Rae province also hosts the uranium deposits of the Athabasca basin. The age of the mineralization is reported to be approximately 1.87 Ga (Halpin 2010).

Fig. 2.12 The location of Hoidas Lake, Saskatchewan, Canada. Modified after Google Maps (2015)

The Hoidas Lake REE deposit is a branching and reuniting system of veins in granitic to tonalitic intrusive rocks. Apatite and allanite carry most of the REE, and only minor amounts of monazite and bastnaesite are present. The main REEs present are La, Ce, Pr, and Nd, with minor Sm and trace Dy. The grade of the deposit is reported to between 2 and 4 % total REE (Halpin 2010).

Concerning the origin of the mineralizations there are different hypotheses, which are summarized by Halpin (2010). It is suggested, that they are related to alkaline or carbonatitic magma.

Exploitation of the Hoidas Lake deposit is planned by the Great Western Minerals Group.

2.4.8 *Thor Lake*

Thor lake is located in the Northwest Territories of Canada, situated 5 km north of the Hearne Channel of the Great Slave Lake, and approximately 100 km east-southeast of Yellowknife. The location is given in Fig. 2.13 (Avalon Rare Metals 2015a).

The Nechalacho deposit (Thor Lake) is located at the Southern Margin of the Archaean Slave Province of the Canadian Shield. It is centered within the Blachford Igneous Complex, which consists of rocks of an early suite of gabbro, quartz syenite and granite, and a later suite of granite, with a core of syenite (Thor Lake syenite). The early suite is relatively aluminous, whereas the rocks of the later suite

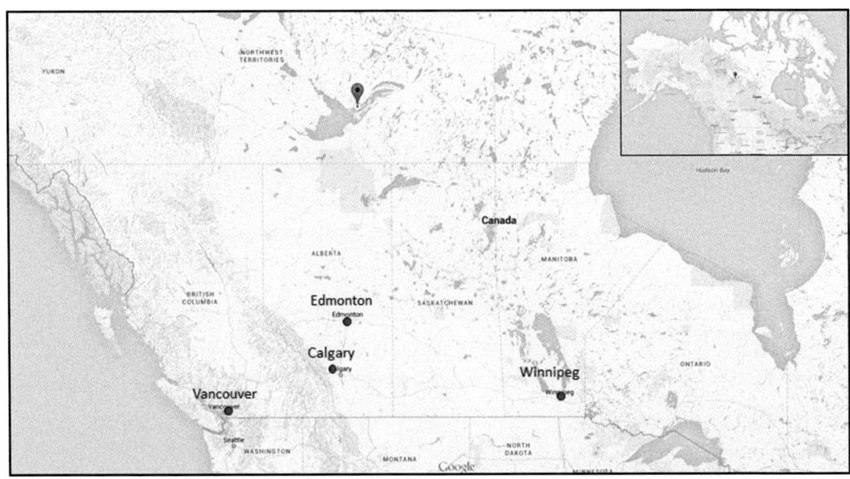

Fig. 2.13 The location of Thor Lake, NWT, Canada. Modified after Google maps (2015)

are of alkaline nature. These rocks also are prominently rich in Nb, REEs, F and partly in Be and Li (Cerny and Trueman 1985).

Avalon Rare Metals Inc., which owns the Nechalacho deposit, has completed a feasibility study on the deposit, which appears to be enriched in the heavy rare earths (HREEs) (Avalon Rare Metals 2015b, c).

The estimated measured mineral resources in the base case now stand at 12.56 million tons averaging 1.71 % TREO,[10] 0.38 % HREO and 22.5 % HREO/TREO (Avalon Rare Metals 2015c).

2.4.9 *Strange Lake and Misery Lake*

2.4.9.1 Strange Lake

The Strange Lake complex (a.k.a. the Lac Brisson complex), on the border of Quebec and Labrador, Canada, is a peralkaline granite which intruded metamorphic rocks and quartz monzonite. It occurs 250 km northeast of Schefferville, Quebec, and 150 km west of Nain, Labrador. The emplacement age of the earliest intrusion is approximately 1240±2 Ma (Miller et al. 1997). The location is given in Fig. 2.14.

The complex consists of three main units, of which the last has rare metal mineralization (Miller et al. 1997). These authors mention that the complex is similar in age, tectonic setting, spatial association (through plate reconstruction) and chemistry to rocks of the Gardar Province in South Greenland (Ilímaussaq, see

[10]TREO = Total Rare Earth Oxide, HREO = Heavy Rare Earth Oxide.

Fig. 2.14 The location of Strange Lake and Misery Lake. Modified after Google Maps (2015)

2.4.4). The average composition of the ore from Strange Lake is given in Table 2.10 (Quest Rare Minerals 2014).

The deposit will be exploited by Quest Rare Minerals, Ltd.

2.4.9.2 Misery Lake

The Misery Lake deposit (Fig. 2.14) was discovered by Quest Rare Minerals Ltd. in 2007 during reconnaissance sampling of an unusual regional magnetic feature (Google Maps 2015). Analyses of bedrock samples gave concentrations of 27 % Fe_2O_3, 1.2 % P_2O_5, 1.5 % TiO_2 and 2.25 % total REE-oxide. Further investigations in 2009 revealed a large REE-bearing alkaline intrusive complex. Misery Lake is located 120 km south of the Strange Lake deposit. It covers a total area of 44,856 ha (Quest Rare Minerals 2014). The intrusion is geologically and geochemically similar to that of Strange Lake.

Table 2.10 Average composition of the ore from Strange Lake (Quest Rare Minerals 2014)

	Enriched zone	Granite domain
Resource	20,020 tons	258,108 tons
Composition	%	%
La_2O_3	0.150	0.120
CeO_2	0.360	0.270
Pr_6O_{11}	0.039	0.030
Nd_2O_3	0.140	0.110
Sm_2O_3	0.036	0.024
Eu_2O_3	0.002	0.001
Gd_2O_3	0.039	0.023
Tb_4O_7	0.009	0.005
Dy_2O_3	0.066	0.032
Ho_2O_3	0.015	0.007
Er_2O_3	0.049	0.022
Tm_2O_3	0.008	0.003
Yb_2O_3	0.051	0.022
Lu_2O_3	0.007	0.003
Y_2O_3	0.470	0.220
Total	1.018	0.451

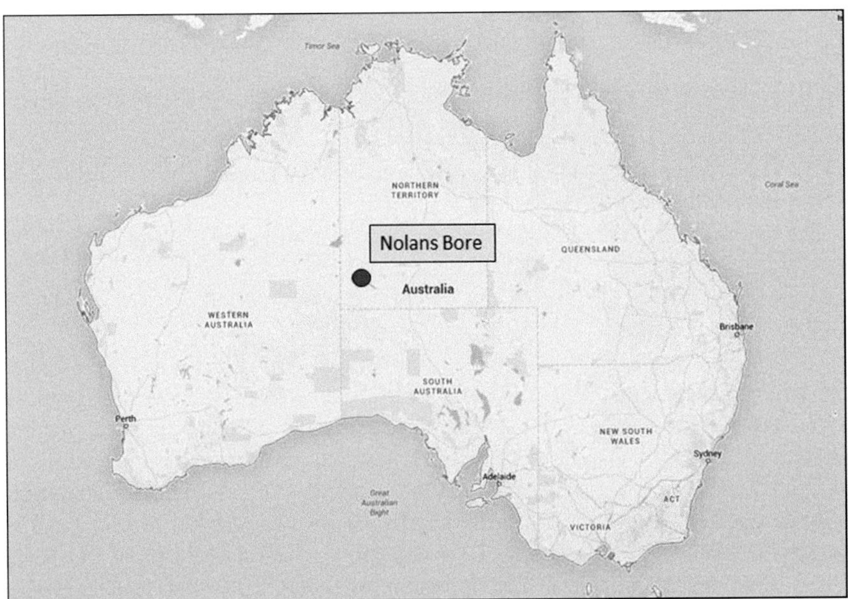

Fig. 2.15 The location of the Nolans Bore deposit, Northern Territory, Australia. Modified after Google Maps (2015)

2.4.10 Nolans Bore Deposit, Australia

The Nolans Bore deposit (Arunta Region, NT) is located in a zone which also contains the Mud Tank Carbonatite, Mordor Igneous Complex, and several tin- and tantalum-bearing pegmatites (Fig. 2.15). The deposit was discovered in 1995 (Arafura Resources 2014).

The Nolans Bore mineral field of carbonatites, pegmatites, and other REE occurrences consist of mineralized fluorapatite veins and breccia zones, which are hosted in majority by granite (metamorphosed to gneiss).

The granite-gneiss has been strongly kaolinized due to weathering. The weathering zone shows secondary enrichment of REE. The average REE-composition is given in Table 2.11.

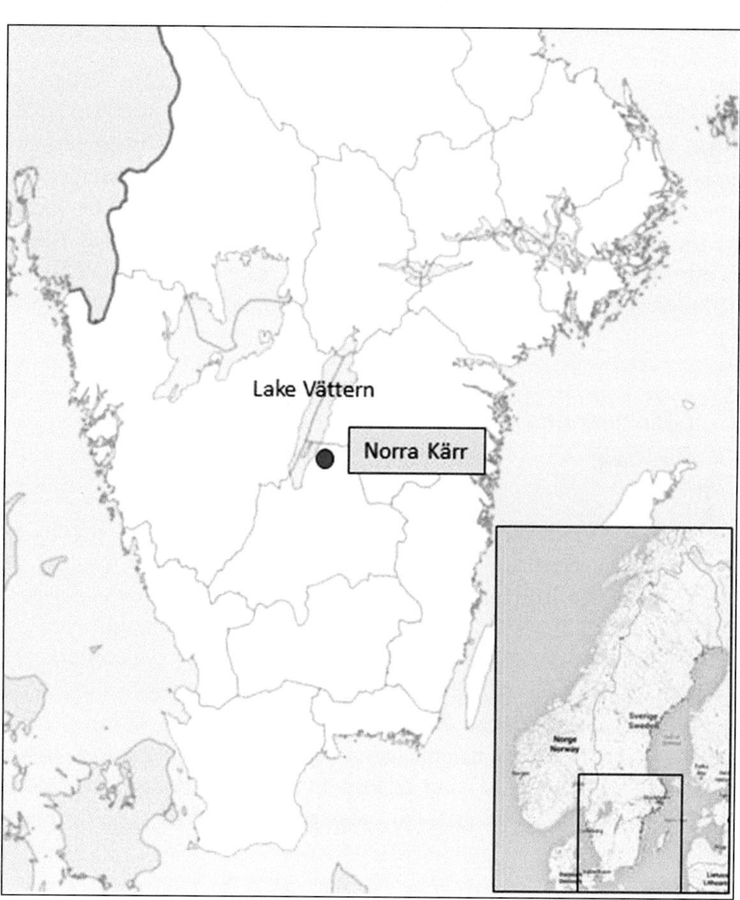

Fig. 2.16 The location of the Norra Kärr deposit. Modified after Google Maps (2015)

Table 2.11 Average REE-composition of the Nolans Bore deposit (Arafura Resources)

Element	% of total REO
La	19.1
Ce	48.7
Pr	5.9
Nd	20.6
Sm	2.3
Eu	0.39
Gd	0.99
Tb	0.08
Dy	0.32
Y	1.35
Other	0.21

2.4.11 Norra Kärr, Sweden

The Norra Kärr alkaline igneous complex is located in southern Sweden, in the province of Småland, about 1.5 km east of Lake Vättern and about 15 km NNE of the small town of Gränna (Fig. 2.16). Discovered by A. E. Törnebohm in 1906 (Törnebohm 1906), the complex was studied in detail by von Eckermann (1968). The intrusion is roughly elliptical in shape (1200 m × 400 m). The age of the intrusion is 1489 ± 8 Ma (Sjöqvist et al. 2013). The deposit is exploited by Tasman Metals Ltd, and it is especially enriched in the HREEs, which make out more than 50 % of the total REE content (Tasman Metals 2014b). The main ore mineral is eudialyte.

2.4.12 Lovozero and Khibina Massifs, Kola Peninsula, Russia

The Kola alkaline province occupies the Kola Peninsula, northern Karelia and the adjoining regions of Northern Finland. Two large agpaitic intrusions are the key magmatic centers. The Khibina[11] massif consists of intercalations of K–Na and K nepheline syenites with typical ultrabasic-alkaline and carbonatitic rocks. In the Lovozero massif, there are agpaitic lujavrites[12] (type locality), which form a layered complex similar to Ilímaussaq, Greenland. Geochronological data show the age to be Paleozoic (Arzamastsev et al. 2008).

The larger of the two agpaitic nepheline syenite intrusions is that at Khibina, with an exposed surface of 1327 km^2. Close by there is a second agpaitic complex with an exposed surface of 650 km^2, the Lovozero intrusion (Fig. 2.17). It is located

[11]Khibina is also spelled as Khibiny, or Khibini.

[12]The name Lujavrite is derived from the Saami-word Lujávri, meaning (Lake) Lovozero.

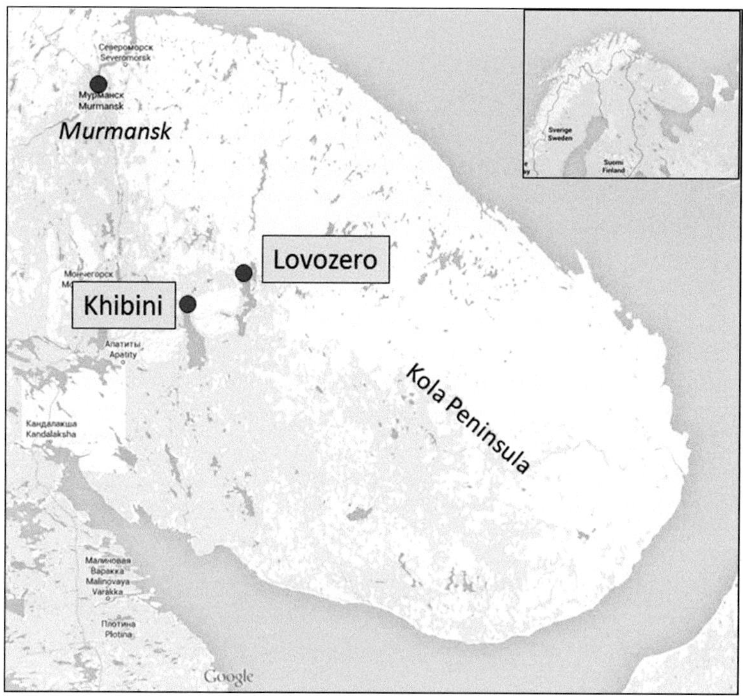

Fig. 2.17 Location of the Lovozero and Khibiny intrusions. Modified after Google maps (2015)

about 20 km north east of Khibina. Both alkaline complexes include apatite and loparite deposits which are of economic importance.

Apatite from the Khibina alkaline complex is mainly fluorapatite $Ca_5(PO_4)F$, with enrichment of light rare earth elements (La, Ce, Pr, Nd, Sm, Eu). The investigated LREE + Y contents range from 4268 to 4464 ppm (with an average of 656 ppm for Nd). The content of the more rare heavy rare earth elements (Gd, Dy, Er, Yb) is minor (HREE = 405 ppm). (Stoltz and Meyer 2012).

2.4.13 *Nkwombwa Hill Carbonatite Deposit, Zambia, and Other East- and Middle-African REE-Deposits*

Although there is more than one deposit in Central Africa, the most important at the moment is the Nkwombwa Hill Carbonatite Deposit, Zambia (Turner et al. 1989). Originally mapped as a limestone deposit, it was later recognised as a carbonatite plug (Zambezi et al.1997). The carbonatite plug is elliptical in form, measuring some 600 × 1200 m and some 300 m in height. The age was determined at 679 ± 25 Ma (Snelling 1965), which is Neoproterozoic.

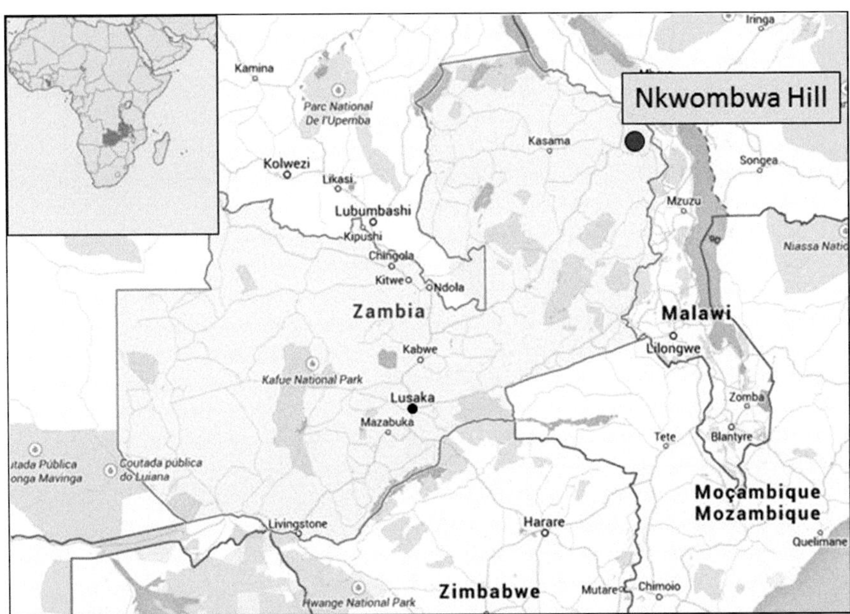

Fig. 2.18 Nkwombwa Hill in NE Zambia. Modified after Google Maps (2015) and Zambezi et al. (1997)

Emplacement is thought to have been controlled by a major NE–SW trending fault. The sills and dykes of ferrocarbonatite contain irregularly distributed angular xenoliths, typically 10 cm or more in size, which locally make up some 20–25 % by volume of the rock.

Rare earth mineralization at Nkwomba Hill is restricted to these xenoliths (Zambezi et al. 1997). The average composition of the xenoliths is given in Table 2.12. The REEs are largely LREEs. The high-grade rare earth mineralization will be exploited by Galileo Resources Plc, a 2005 founded, UK registered company.

Other carbonatite-related REE-deposits in Central and East Africa are, for instance:

- The Kangankunde Carbonatite Complex, Malawi. This is one of several carbonatite complexes in southern Malawi associated with the Shire Valley section of the East African Rift System. The Kangankunde Carbonatite Complex differs significantly from other carbonatite in the marked absence of nepheline syenite and other silicate rocks like lamprohyres and melanephelinite (Duraiswami and Shaikh 2014). Main REE-minerals are monazite-Ce and bastnaesite-Ce, and to a lesser extent florencite-goyazite[13] (Wall and Mariano 1996). Resources at the

[13]Goyazite is: $SrAl_3(PO_4)_2(OH)_5$
Florencite is: $(Ce, La)Al_3(PO_4)_2(OH)_6$

Table 2.12 Average composition of REE-rich xenoliths from Nkwombwa Hill, Zambia (Zambezi et al. 1997)

	wt%
SiO_2	0.52
Fe_2O_3	1.09
La_2O_3	8.25
Ce_2O_3	15.35
Pr_6O_{11}	1.25
Nd_2O_3	1.76
Sm_2O_3	0.15
Eu_2O_3	0.12
Gd_2O_3	0.26
SrO	1.95
BaO	12.28
MnO	0.58
CaO	15.65
MgO	7.70
P_2O_5	1.27
LOI	28.10
Total	96.30

Kangankunde deposit are estimated to be 2.53 Mt at a grade of 4.24 % rare earth minerals (Yager 2011).

- The Tundulu Complex, Malawi, is a carbonatite with veins with mainly LREE mineralization. The REE-minerals are synchysite, parisite, and bastnaesite (Ngwenya 1994).
- The Songwe Hill Rare Earth Element (REE) Project, which will be exploited by Mkango Resources Ltd., is located within the 100 %-owned Phalombe License, which covers a portion of the Cretaceous Chilwa Alkaline Province in Southern Malawi (Mkango Resources Ltd 2014).

2.4.14 Maoniuping, Sichuan, China

In Sichuan, China, the Maoniuping Deposit is the second largest LREE deposit in China (Fig. 2.19). It occurs in the northern Jinpingshan Mountains, a Cenozoic intracontinental orogenic belt.

It is a vein type deposit, which is hosted within, and is genetically related to, a carbonatitic alkaline complex. The deposit is very similar to the Californian Mountain Pass deposit. The similarities are the high concentrations of bastnaesite and barite, low niobium content, and the presence of sulphides (Wang et al. 2001). It has proven reserves of 0.4 million tons of ore grading at 2 % REE oxides. The discovery of the deposit was described by Pu (1988). The age of the carbonatites related to the deposit is 27.8–40.3 Ma (Cheng et al. 2003; Mindat.Org 2015).

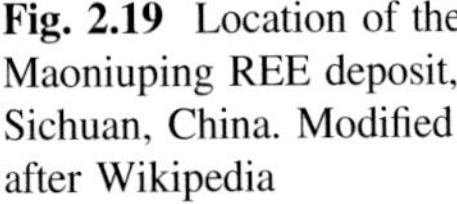

Fig. 2.19 Location of the Maoniuping REE deposit, Sichuan, China. Modified after Wikipedia

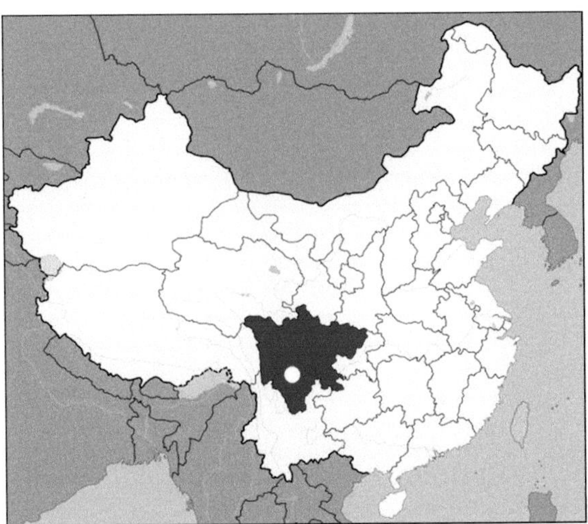

2.4.15 Chinese Ion Adsorption Deposits

China's weathered-crust elution-deposit rare earth ore, called ion-adsorption rare earth ore, is unique. This type of ore was discovered in 1969 in the Jiangxi Province. There are two types of deposits: light REE type (A) and heavy REE type (B). It was found also in other Chinese provinces: Fujian, Hunan, Guangdong, and Guanxi. The provinces where these deposits are found are given in Fig. 2.20. The

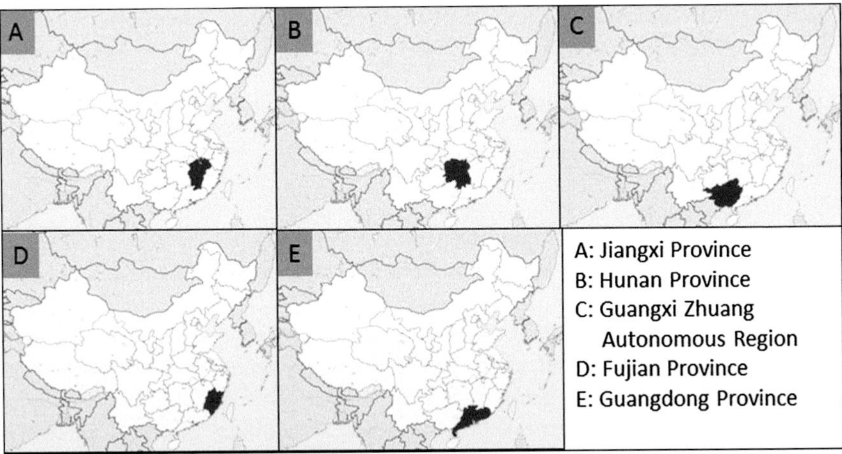

Fig. 2.20 The provinces of China where the Ion Adsorption Ores have been found. Image composed and modified after Wikipedia

Table 2.13 Average compositions of two types of REE ion adsorption ores from China (Zhi Li and Yang 2014)

Oxide	wt%	wt%
	Type A	Type B
La_2O_3	31–40	2–5
CeO_2	3–7	1–5
Pr_6O_{11}	7–11	1–2
Nd_2O_3	26–35	3–5
Sm_2O_3	4–6	2–4
Eu_2O_3	0.5	0.12
Gd_2O_3	4	6
$(Tb–Lu)_2O_3$	4–5	15–20
Y_2O_3	9–11	>60

deposits are considered to have formed by weathering of granite and effusive rocks over many years. REEs are in the form of positive hydrated ions, adsorbed on the surface of clay minerals like kaolinite, halloysite and illite. The ores are relatively low grade, generally 0.05–0.5 % REO, but with high amounts of HREE (Zhi Li and Yang 2014). See Table 2.13 for average compositions. The known reserve of these ore deposits is more than 1×10^6 tons REO (rare earth oxide). The known reserve of HREE in China is 80 % of the total world reserve of REE (Zhi Li and Yang 2014).

2.4.16 *Dong Pao, Vietnam*

In the north-western part of the Lai Châu Province, in the Tam Duong district of Vietnam, a major rare earth element deposit occurs (Fig. 2.21). This deposit, called Dong Pao, consists of irregular shapes with nests, lenses and veins in a shear zone of limestone that was hydrothermally altered. Mineralization consists of bastnaesite, fluorspar and parisite as the main ore minerals. The deposits in the Nam Xe area are similar to the Dong Pao deposit (Fong Sam 2011).

The Dong Pao Rare Earth mine covers a total area of 11 square miles, and has estimated reserves of 5 million tons (Fong Sam 2011).

The Dong Pao Mine is jointly operated by Lai Chau-VIMICO Rare Earth Joint Stock Co. and the Japanese Dong Pao Rare Earth Development Company (Talk Vietnam 2012).

Fig. 2.21 Location of the Dong Pao Rare Earth Deposit, Vietnam. After Google Maps (2015), and Wikipedia (2015a–f)

2.5 Resources in the Deep Sea

In recent years more and more research has been conducted to locate ore deposits in the deep sea. Although not exclusively a target, REEs have also been found in deep sea deposits. The origin of the REE may be diverse: rivers, aeolian processes and hydrothermal processes (de Baar et al. 1985; Ederfeld and Greaves 1982).

2.5.1 *Pelagic Muds*

Kato et al. (2011) found vast amounts of deep sea mud in the Pacific Ocean. The muds are generally metalliferous sediment, zeolitic clay, and pelagic[14] red clay. Thicknesses vary from a few meters to 70 m. They occur up to 50 m below the sea floor. These deposits occur in the eastern South Pacific and central North Pacific. They contain 1000–2300 ppm total REY (REE + Y). In the eastern South Pacific (5°–20° S, 90°–150° W), the REY-rich mud has high REY contents, 1000–2230 ppm total REY and 200–430 ppm total HREE. REY contents of the mud are comparable to or greater than those of the southern China ion-absorption-type deposits (where ΣREY = 500–2000 ppm; ΣHREE = 50–200 ppm); notably, the HREEs are in general nearly twice as abundant as in the Chinese deposits (Kato et al. 2011).

2.5.2 *Crusts on Seamounts*

Seamounts may have a diverse origin. In general, any conical or steep volcanic feature at the sea floor is referred to as a seamount, and these may be, or may not be, volcanically active. Seamounts occur in groups or in chains. They originate from different eras, and are mostly found at convergent plate boundaries and in areas of vertical tectonic movement. They also occur at intersections of ridge faults[15] and transform faults,[16] at spreading centers and at hotspots. Seamounts range from small domes of tens of meters to large structures of several kilometers in height. Commonly, they have steep outer slopes, flat or nearly flat circular summit areas

[14]Pelagic red clay: red colored fine-grained sediment that accumulates as the result of the settling of particles to the floor of the open ocean, far from land. The color results from coatings of iron oxide and manganese oxide on the sediment particles (Source Wikipedia (2015a) Pelagic Red Clay). The word "pelagic" comes from the Greek, and means "open sea".

[15]Ridge faults are central faults from mid-oceanic ridges, which are huge submarine mountain chains.

[16]Transform faults are also known as conservative plate boundaries, as they neither create nor destroy lithosphere. These faults have a relative motion which is predominantly horizontal. They tend to be approximately at right angles to mid-oceanic ridges.

and collapse features (calderas, pits, craters). Metal deposits are usually found on the slopes and flanks of the seamounts.

Cerium was reported among others by Baturin and Yushina (2007). In a publication from 2008, the ISA confirmed the occurrence of cobalt-rich crusts (ISA, 2008).

2.5.3 Exploitation of Deep-Sea Deposits

The main problem with exploitation of deep-sea deposits is their depth: they tend to occur in water which is several kilometers deep. It is evident that this entails substantial technical difficulties. The exploitation of these deposits, outside the Exclusive Economic Zone (EEZ) of coastal and island states (200 miles offshore) is under the supervision of the International Seabed Authority (ISA 2015).

References

Andreoli MAG, Smith CB, Watkeys M, Moore JM, Ashwal LD, Hart RJ (1994) The geology of the Steenkampskraal monazite deposit, South Africa: implications for REE-Th-Cu mineralization in charnockite-granulite terrains. Econ Geol 89:994–1016

Anthony JW, Bideaux RA, Bladh KW, Nichols MC (eds) Handbook of mineralogy, online version, Mineralogical Society of America. http://www.handbookofmineralogy.org/. Accessed Sept 2014

Arafura Resources (2014) http://www.arultd.com/our-projects/nolans/rare-earths-mix.html. Accessed Oct 2014

Arzamastsev A,Yakovenchuk V, Pakhomovsky Y, Ivanyuk G (2008) The Khibina and Lovozero alkaline massifs: geology and unique mineralization. IGC excursion No 47 (excursion guide)

Avalon Rare Metals (2015a) http://www.avalonraremetals.com/_resources/factsheet/ProjectSheet.pdf. Accessed Aug 2015

Avalon Rare Metals (2015b) Nechalacho Rare Earth Elements ("REE") Project (http://www.avalonraremetals.com/_resources/factsheet/ProjectSheet.pdf). Accessed Aug 2015

Avalon Rare Metals (2015c) Nechalacho project: resources and reserves. http://avalonraremetals.com/nechalacho/resources_reserves. Accessed Aug 2015

Baturin GN, Yushina IG (2007) Rare earth elements in phosphate-ferromanganese crusts on Pacific seamounts. Lithol Min Resour 42(2):101–117

Berzelius JJ (1824) Undersökning af några Mineralier. 1. Phosphorsyrad Ytterjord. Kongliga Svenska Vetenskaps-Akademiens Handlingar 2:334–338

Berzelius JJ (1825) Account of two newly discovered mineral species. Edinb J Sci 3:327–332

Blaxland A, Van Breemen O, Steenfelt A (1976) Age and origin of agpaitic magmatism at Ilímaussaq, South Greenland: Rb-Sr study. Lithos 9:31–38

Breithaupt (1829) Über den Monazit, eine neue Specie des Mineral Reichs. J für Chemie und Physik, 55:301–303

Castor SB (2008) The Mountain Pass rare-earth carbonatite and associated ultrapotassic rocks, California. Can Mineral 46:779–806

Cerny P, Trueman DL (1985) Polylithionite from the rare-metal deposits of the Blachford Lake alkaline complex, N.W.T, Canada. Am. Mineral 70:1127–1134

Chao ECT, Back JM, Minkin JA, Tatsumoto M, Wang J, Conrad JE, McKee EH, Zonglin H, Qingrun M (1997) The sedimentary carbonate-hosted giant Bayan Obo REE-Fe-Nb ore deposit of Inner Mongolia, China: a cornerstone example for giant polymetallic ore deposits of hydrothermal origin. USGS Bull. 2143, 65 p

Cheng X, Huang Z, Liu C, Qi L, Li W, Guan T (2003) Geochemistry of carbonatites in Maoniuping REE deposit, Sichuan province, China. Sci China, Ser D. 46(3):246–256

de Baar HJW, Bacon MP, Brewer PG (1985) Rare earth elements in the Pacific and Atlantic Oceans. Geochim Cosmochim Acta 49:1943–1959

Deer WA, Howie RA, Zussman J (1986a) Eudialyte-Eucolite. The rockforming minerals, volume 1B, disilicates and ringsilicates, 2nd edn. Longman Scientific and Technical, Harlow, pp 348–363

Deer WA, Howie RA, Zussman J (1986b) Allanite. The rockforming minerals, volume 1B, disilicates and ringsilicates, 2nd edn. The Geological Society, Bath, United Kingdom, pp 151–179

Deer WA, Howie RA, Zussman J (2013) An introduction to the rock-forming minerals, 3rd edn. The Mineralogical Society, London, p 478 (Monazite), p 63, (Allanite)

Desharnais G, Camus Y, Bisaillon B (2014) Resources for the Tantalus rare earth ionic clay project, Northern Madagascar. SGS Canada Inc., NI 43-101 Technical Report. 165 p

Drew LJ, Qingrun M, Weijun S (1990) The Bayan Obo iron-rare-earth-niobium deposits, Inner Mongolia, China. Lithos 26:43–65

Duraiswami RA, Shaikh TN (2014) Fluid-rock interaction in the Kangankunde Carbonatite Complex, Malawi: SEM based evidence for late stage pervasive hydrothermal mineralisation. Cent Eur J Geosci 6(4):476–491

Ederfeld H, Greaves MJ (1982) The rare earth elements in seawater. Nature 296:214–219

Encyclopedia Brittanica http://www.britannica.com/EBchecked/topic/438414/Bayan-Obo. Accessed Sept 2014

Finch AA, Goodenough KM, Salmon HM, Andersen T (2001) The petrology and petrogenesis of the North Motzfeldt Centre, Gardar Province, South Greenland. Mineral Mag 65(6):759–774

Förster HJ (1998a) The chemical composition of REE-Y-Th-U-rich accessory minerals in peraluminous granites of the Erzgebirge-Fichtelgebirge region, Germany. Part I: the monazite-(Ce)-brabantite solid solution series. Am Mineral 83:259–272

Förster HJ (1998b) The chemical composition of REE-Y-Th-U-rich accessory minerals in peraluminous granites of the Erzgebirge-Fichtelgebirge region, Germany. Part II: Xenotime Am Mineral 83:1302–1315

Fitton JG, Upton BGJ (1987) Introduction to Alkaline Igneous rocks. In: Alkaline Igneous rocks. Geol Soc Spec Publ 30:ix–xiv

Fong Sam Y (2011) The mineral industry of Vietnam, USGS 2011 minerals yearbook. 28.1–28.14

GEUS (2011) Minex, greenland mineral exploration newsletter. 40:8 p

Google Maps (2015) (https://maps.google.com)

Gratz R, Heinrich W (1997) Monazite-xenotime thermobarometry: experimental calibration of the miscibility gap in the binary system $CePO_4$-YPO_4. Am Mineral 82:772–780

Greenland Minerals and Energy (http://www.ggg.gl/projects/specialty-metals-kvanefjeld). Accessed Oct 2014

Gupta CK, Krishnamurthy N (2005) Extractive metallurgy of the rare earths. CRC Press, Boca Raton 484 p

Halpin KM (2010) The characteristics and origin of the Hoidas Lake REE deposit. M.Sc. Thesis, University of Saskatchewan, 257 p

Handbook of Mineralogy (2001) Eudialyte, Mineral Data Publishing, version 1.2

Harris C, Cressey G, Bell ID, Atkins FB, Beswetherick S (1982) An occurrence of rare-earth-rich eudialyte from Ascension Island, South Atlantic. Mineral Mag 46:421–425

Hewett DF (1954) History of discovery at Mountain Pass, California. In: Rare-earth mineral deposits of the Mountain Pass District, San Bernardino, County, California. USGS Prof paper 261: iii–vi

Hisinger W (1838) Analyser af några svenska mineralier. 2.Basiskt Fluor-Cerium från Bastnäs. Kongl. Vetenskaps-Akademiens Förhandlingar 187–1891 (as Basiskfluor-cerium)

Hoatson DM, Jaireth S, Miezitis Y (2011) The major rare-earth-element deposits of Australia: geological setting, exploration, and resources. Geosci Aust 204 p

International Seabed Authority (ISA) https://www.isa.org.jm. Accessed July 2015

Jones I, Hancox PJ (2012) Steenkampskraal rare earth element project South Africa. Technical report and resource estimate. Great Western Minerals Group, Ltd. 145 p

Kato Y, Fujinaga K, Nakamura K, Takaya Y, Kitamura K, Ohta J, Toda R, Nakashima T, Iwamori H (2011) Deep-sea mud in the Pacific Ocean as a potential resource for rare-earth elements. Nature Geosci 4:535–539

Klaproth MH (1810) Chemische Untersuchung des rothen Granats aus Grönland, Beiträge zur chemischen Kenntniss der Mineralkörper, 5, Rottmann Berlin, pp 131–137

Lottermoser BG (1990) Rare-earth element mineralisation within the Mount Weld carbonatite laterite. Western Australia Lithos 24:151–167

Lynas Corporation (2013) Quaterly report for the period ending 31 Dec 2012

McBirney AR (1993) Igneous petrology, 2nd edn. Jones and Bartlett Publishers, Boston, London, pp 446–450

Miller RR, Heamen LM, Birkett TC (1997) U-Pb zircon age of the Strange Lake peralkaline complex: implications for Mesoproterozoic peralkaline magmatism in north-central Labrador. Precambrian Res. 81:67–82

Mindat.org: Bayan Obo (http://www.mindat.org/loc-720.html). Accessed Sept 2014; Monazite (http://www.mindat.org/min-2751.html). Accessed Aug 2014; Eudialyte (http://www.mindat.org/min-1420.html). Accessed Aug 2014; Bastnaesite (http://www.mindat.org/min-560.html). Accessed Aug 2014

Mindat.org: Maoniuping (http://www.mindat.org/loc-73232.html). Accessed July 2015

Mkango Resources Ltd (2014) (http://www.mkango.ca/s/songwe.asp). Accessed Dec 2014

Molycorp (2014) (www.molycorp.com). Accessed Aug 2014

Ngwenya BT (1994) Hydrothermal rare earth mineralisation in carbonatites of the Tundulu complex, Malawi: processes at the fluid/rock interface. Geochim Cosmochim Acta 58 (9):2061–2072

Pu G (1988) Discovery of an alkali-pegmatite carbonatite complex zone in Maoniuping, south-western Sichuan Province. Geol Rev 34(1):86–92 (In Chinese, with English Abstract)

Olson JC, Shaw DR, Pray LC, Sharp WN (1954) Rare-earth mineral deposits of the Mountain Pass District, San Bernardino County, California. USGS Prof Paper 261, 75 p

Orris GJ, Grauch RI (2002) Rare earth element mines, deposits, and occurrences. USGS open file report, 02–189, 174 p

Qiao X, Gao L, Peng Y, Zhang Y (1997) Composite stratigraphy of the Sailinhudong group and ore-bearing Micrite Mounds in the Bayan Obo Deposits, Inner Mongolia, China. Acta Geol Sinica 71(4):357–369

Quest Rare Minerals (2014) Misery Lake rare earth project. (http://www.questrareminerals.com/misery_lake.php). Accessed Nov 2014

Ribeiro Olivo G, Williams-Jones AE (1999) Hydrothermal REE-rich eudialyte from the Pilanesberg complex, South Africa. Can Mineral 37:653–663

Snelling NJ (1965) Age determination of three African carbonatites. Nature 205:492

Sjöqvist ASL, Cornell DH, Andersen T, Erambert M, Ek M, Leijd M (2013) Three compositional varieties of rare-earth element ore: eudialyte-group minerals from the Norra Kärr Alkaline Complex, Southern Sweden. Minerals 3:94–120

Sörensen H (1974) The alkaline rocks, 1st edn. Wiley, Hoboken, 634 p

Sörensen H (1992) Agpaitic nepheline syenites: a potential source of rare elements. App Geochem 7:417–427

Sörensen H (1997) The agpaitic rocks—an overview. Mineral Mag 61:485–498

Sörensen H (ed) (2001) The llimaussaq alkaline complex, South Greenland: status of mineralogical research with new results. Geol Survey Greenland Bull 190:167 p

Sørensen LL, Kalvig P (2011) The rare earth element potential in Greenland. Geolog Survey Denmark Greenland (GEUS) 12 p

Steenfelt A (2012) Rare earth elements in Greenland: known and new targets identified and characterised by regional stream sediment data. Geochem: Expl Environ Anal 12:313–326

Stoltz NB, Meyer FM (2012) Economic potential of rare earth elements in apatite of the Khibina Alkaline complex, Kola Peninsula, Russia. In: 4th International Geologica Belgica Meeting, 2012

Streckeisen A (1967) Classification and nomenclature of igneous rocks. N Jb Miner Abh 107:144–240

Streckeisen A (1980) Classification and nomenclature of volcanic rocks, lamprophyres, carbonatites and melilitic rocks. Geol Rundschau 69:194–207

Talk Vietnam (2012) May 18, 2012 by Vietnamnews. http://tanbreez.com/en/project-overview/tanbreez-elements/?page=1

Tanbreez (2014) http://tanbreez.com/en/project-overview/tanbreez-%e2%80%93-what-is-it/. and http://tanbreez.com/en/project-overview/tanbreez-elements/?page=1 Accessed Oct 2014 and June 2015

Tasman Metals Ltd. (2014a) (http://www.tasmanmetals.com/s/OresMinerals.asp). Accessed Aug 2014

Tasman Metals Ltd. (2014b) (http://www.tasmanmetals.com/s/Norra-Karr.asp). Accessed Nov 2014

Törnebohm AE (1906) Katapleiit-syenit, en nyupptäckt varietet af nefelinsyenit i Sverige [in Swedish]. Swed Geolog Surv (SGU) Ser C 199:1–54

Thomson T (1810) Experiments on allanite, a new mineral from Greenland. Trans Royal Soc Edinburgh 8:371–386

Tukiainen T (2014) The Motzfeld of the Igaliko nepheline syenite complex, South Greenland—a major resource of REE elements. In: ERES2014, proceedings of the 1st European rare earth resources conference, Milos Greece, 04–07 Sept 2014:317–324

Turner DC, Andersen LS, Punukollu SN, Sliwa A, Tembo F (1989) Igneous phosphate resources in Zambia. In: Notholt AJG, Sheldon RP, Davidson DF (eds.) Phosphate deposits of the world 2:247–257

von Eckermann H (1968) New contributions to the interpretation of the genesis of the Norra Kärr alkaline body in Southern Sweden. Lithos 1(1):76–88

van Emden B, Thornber MR, Graham J, Lincoln FJ (1997) The incorporation of actinides in monazite and xenotime from placer deposits in Western Australia. Can Mineral 35:95–104

Visser WA (1980) Geological nomenclature. Royal Geol Mining Soc Netherlands 539 p

Wall F, Mariano AN (1996) Rare earth minerals in carbonatites—a discussion centred on the Kangankunde carbonatite, Malawi. In Jones AP, Wall F, Williams CT (eds) Rare earth minerals—chemistry, origin and ore deposits. The Mineralogical Society, Chapman and Hall, London, United Kingdom, Series 7:193–225

Wang J, Tatsumoto M, Li X, Premo WR, Chao ET (1994) A precise ^{232}Th-^{208}Pb chronology of fine grained monazite: age of the Bayan Obo, REE-Fe-Nb ore deposit China. Geochim Cosmochim Acta 58(15):3155–3169

Wang D, Yang J, Yan S, Xu J, Chen Y, Pu G, Luo Y (2001) A special orogenic-type rare earth element deposit in Maoniuping, Sichuan, China: geology and geochemistry. Resour Geol 51 (3):177–188

Watt GR (1995) High-thorium monazite-(Ce) formed during disequilibrium melting of metapelites under granulite-faciës conditions. Mineral Mag 59:735–743

Webmineral: Monazite (http://webmineral.com/data/Monazite-(Ce).shtml#.Vmqg4U2FOUk). Accessed Aug 2014; Xenotime (http://webmineral.com/data/Xenotime-(Y).shtml#.VmqgtU2FOUk). Accessed Aug 2014; Eudialyte (http://webmineral.com/data/Eudialyte.shtml#.VFc7NqNgWcw). Accessed Nov 2014

Wikipedia, Binnen-Mongolië (http://nl.wikipedia.org/wiki/Binnen-Mongoli%C3%AB). Accessed Aug 2014

Wikipedia (2015a) Pelagic Red Clay https://en.wikipedia.org/wiki/Pelagic_red_clay. Accessed June 2015

Wikipedia (2015b) Jiangxi (https://en.wikipedia.org/wiki/Jiangxi)

Wikipedia (2015c) Gunagxi Zhuang (https://en.wikipedia.org/wiki/Guangxi)

Wikipedia (2015d) Hunan (https://en.wikipedia.org/wiki/Hunan)

Wikipedia (2015e) Fujian (https://en.wikipedia.org/wiki/Fujian)

Wikipedia (2015f) Guangdong (https://en.wikipedia.org/wiki/Guangdong)

Willett GC, Duncan RK, Rankin RA (1989). Geology and economic evaluation of the Mount Weld carbonatite, Laverton, Western Australia. In: Kimberlites and related rocks. GSA Special Publication 14, Blackwell Scientific Publications: 1215–35

Yager TR (2011) The mineral industry of Malawi. In: 2011 Minerals Yearbook. USGS: 27.1–27.4

Yang X-M, Le Bas MJ (2004) Chemical compositions of carbonate minerals from Bayan Obo, Inner Mongolia, China: implications for petrogenesis. Lithos 72:97–116

Yang XY, Sun WD, Zhang YX, Zheng Y-Z (2009) Geochemical constraints on the genesis of the Bayan Obo Fe–Nb–REE deposit in Inner Mongolia, China. Geochim Cosmochim Acta 73:1417–1435

Zaitsev AN, Williams CT, Jeffries TE, Strekopytov S, Moutte J, Ivashchenkova OV, Spratt J, Petrov SV, Wall F, Seltmann R, Borozdin AP (2014a) Rare earth elements in phoscorites and carbonatites of the Devonian Kola Alkaline province, Russia: Examples from Kovdor, Khibina, Vuoriyarvi and Turiy Mys complexes. Ore Geol Rev 61:204–225

Zaitsev AN, Wall F, Chakhmouradian AR (2014b) Rare earth element minerals in carbonatites of the Kola Alkaline Province (Northern Fennoscandia). In: Proceedings of ERES2014: 1st European rare earth resources conference, Milos, Greece, 04–07 Sept 2014:343–347

Zhang P, Kejie T, Yang Z, Yang X, Song R (2002) Rare earths, niobium and tantalum minerals in Bayan Obo ore deposit and discussion on their genesis. J Rare Earths 20(2):81–86

Zambezi P, Voncken JHL, Hale M, Touret JLR (1997) Bastnaesite-(Ce) at the Nkombwa Hill carbonatite complex, Isoka District, Northeast Zambia. Mineral Petrol 59:239–250

Zhi Li L, Yang X (2014) China's rare earth ore deposits and beneficiation techniques. In: Proceedings of ERES2014: 1st European rare earth resources conference, Milos, 04–07 Sept 2014 pp 26–36

Chapter 3
Physical and Chemical Properties of the Rare Earths

Abstract This chapter discusses the chemical and physical properties of the lanthanides, some of which are in a certain way peculiar. It discusses the oxidation states of the REE, and the phenomenon called the **lanthanide contraction** (meaning that the atomic radius decreases with increasing atomic number in the series lanthanum–lutetium). It lists the isotopes known per element, and explains the radioactivity of promethium, the only element of the rare earths that has only radioactive isotopes and no stable isotopes. Magnetism and luminescence also are discussed.

3.1 Introduction

The rare earths are divided into the lanthanide group, and the elements scandium and yttrium. The lanthanides constitute a special group of elements that have an atomic structure different from the other elements, although they are somewhat akin to the actinide series. The lanthanides make up a group of elements ranging from atomic number 57 (lanthanum) to 71 (lutetium). They are all very similar to lanthanum, which is the reason for the name lanthanide. The other two rare earths, scandium and yttrium, are somewhat apart from the lanthanide series, and will be treated separately.

The similarity in characteristics of the lanthanides series includes:

- Similarity in physical properties throughout the series
- In crystalline compounds, they usually have the 3+ oxidation state, although some can also have a 2+ or 4+ oxidation state (Haire and Eyring 1994).
- Coordination numbers[1] in compounds are usually greater than 6.
- Across the series, the coordination number decreases.

[1]Coordination number: this is in chemistry the number of the near neighbors of a central atom in a molecule.

J.H.L. Voncken, *The Rare Earth Elements*, SpringerBriefs in Earth Sciences,
DOI 10.1007/978-3-319-26809-5_3

- They bind preferably with the strong electronegative elements such as oxygen or fluorine.

Common properties of the rare earths.

- The rare earths are silver, silvery-white, or grey metals.
- The metals have a high luster, but tarnish readily in air.
- The metals have high electrical conductivity.
- There are very small differences in solubility and complex formation between the rare earths.
- The rare earth metals naturally occur together in minerals.
- Rare earths are found with non-metals, usually in the 3+ oxidation state.

The similarity may be explained by the electronic configuration of the atoms. This is discussed later.

3.2 The Lanthanide Series

Theoretically, an atom may have the following electron shells: K, L, M, N, O, P, and Q; or 1, 2, 3, 4, 5, 6, and 7; where K (or 1) is the most inward and P (or 7) is the most outward. The electrons in the outer shells are most important for determining the chemical behavior, and how they behave as conductors.

Every shell has also subshells. These are composed of atomic orbitals. The electron subshells of an atom are labelled s, p, d, and f, which is short for: sharp, principal, diffuse, and fundamental (Jensen 2007). The first shell (K) has one subshell, called 1s. The second shell (L) has two subshells called 2s and 2p. The third shell (M) has 3 subshells (3s, 3p, 3d). The fourth shell (N) has 4 subshells (4s, 4p, 4d, 4f).

The numbers of possible electrons per orbital is given in Table 3.1.

The order in which these orbitals are filled is given by the n + ℓ rule (also known as the **Madelung rule** (Madelung 1943).

The **Madelung Rule**[2] is also called the **Aufbau principle** (from German "*Aufbau*" meaning *"build-up, fabric, structure"*). This describes electron configuration and the filling of atomic orbitals.

The rule states:

1. Energy increases with increasing $n + \ell$, where n = the principal quantum number, and ℓ = the azimuthal quantum number (orbital angular momentum quantum number).
2. For identical values of $n + \ell$, energy increases with increasing n.

Resulting from this, the following order of filling of orbitals (from left to right) emerges:

---

[2]Named after Erwin Madelung (1881–1972), German physicist.

Table 3.1 Filling of the subshells

Subshell	Number of orbitals	Total number of possible electrons in each orbital
s	1	2
p	3	6
d	5	10
f	7	14

1s*, *2s*, *2p*, *3s*, *3p*, *4s*, *3d*, *4p*, *5s*, *4d*, *5p*, *6s*, *4f*, *5d*, *6p*, *7s*, *5f*, *6d*, *7p. Madelung's Rule is graphically depicted in Fig. 3.1:

This electronic configuration results in the following effects (Fig. 3.1). For the majority of the neutral lanthanide atoms, the electronic configuration can be given as $4f^{n+1,}$ $5s^2$, $5p^6$, $6s^2$. Loss of one 4f electron and two $6s^2$ electrons leads to the *characteristic 3+ cations.* In some cases, the energy difference between the 4f and 5d electrons is so small, that a 4f electron can be promoted to the 5d orbital. As a result, tetravalent ions form (e.g. Ce^{4+}). Also a 5d-electron may be transferred to a 4f orbital, which results in the formation of divalent atoms, for instance, Sm^{2+}, Eu^{2+}, and Yb^{2+} (Choppin and Rizkalla 1994).

The possible oxidation states of the REEs are listed in Table 3.2.

As previously stated, the REEs are often divided into the light rare earth elements (LREEs) and the heavy rare earth elements (HREEs). This definition is based on the electron configuration of each rare earth element. The LREEs are defined as lanthanum, with atomic number 57 to gadolinium, with atomic number 64. The HREEs are defined as terbium with atomic number 65 to lutetium with atomic number 71, and also includes yttrium, with atomic number 39. The HREEs differ from the LREEs in that they have "paired electrons" (clockwise and

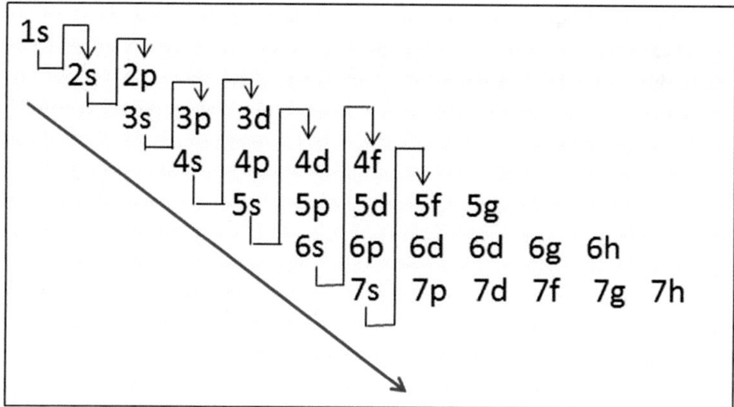

Fig. 3.1 Graphical depiction of the **Madelung Rule**, redrawn after (https://thespectrumofriemannium.wordpress.com/?s=Madelung). "Riemannium" is a fictional element, to describe the Madelung rule

Table 3.2 The oxidation states of the lanthanides

La	Ce	Pr	Nd	Pm	Sm	Eu	Gd	Tb	Dy	Ho	Er	Tm	Yb	Lu
		(2)	(2)		2	2			(2)	(2)		(2)	2	
3	$\underline{3}$	$\underline{3}$	$\underline{3}$	$\underline{3}$	$\underline{3}$	$\underline{3}$	$\underline{3}$	$\underline{3}$	$\underline{3}$	$\underline{3}$	$\underline{3}$	$\underline{3}$	$\underline{3}$	$\underline{3}$
	4	4	(4)					4	(4)					

Oxidation states in parentheses are unstable. The most common oxidation states are underlined (Choppin and Rizkalla 1994)

counter-clockwise spinning electrons). Yttrium is included in the HREE because of similarities in its ionic radius and similar chemical properties. Scandium is sufficiently different in properties to be not classified as either a LREE or a HREE, however, it is often (and also here) discussed together with REE (e.g. Beaudry and Geschneider 1978). See also Sect. 3.2.2. Sometimes also a medium group is identified, consisting of the elements samarium to dysprosium (Gupta and Krishnamurthy 2005).

It is observed that the abundance decreases with increasing atomic number Z: HREEs are much less abundant than LREEs. What's more, according to the Oddo-Harkins Rule,[3] elements with an even atomic number are more abundant than elements with an odd atomic number. For instance, cerium (Z = 58) is more abundant than lanthanum (Z = 57) and praseodymium (Z = 59). Cerium is the dominating rare earth in the LREE, whereas Y is the dominant rare earth in the HREE (Binnemans et al. 2013).

3.2.1 *Lanthanide Oxides*

For most lanthanides, the 3+ oxidation state is the most stable, and therefore almost all REE-oxides are presented as REE_2O_3. However, some of the lanthanides may have several valences in one and the same oxide, so formulas are given to express this phenomenon. Praseodymium oxide usually contains 3+ and 4+ praseodymium in a somewhat variable ratio, depending upon the conditions of formation. Its formula is rendered as Pr_6O_{11}. Similarly, Tb_4O_7, one of the main commercial terbium compounds, contains some Tb_{4+} along with the more stable Tb_{3+}. Ce has the 3+ state as most stable oxidation state, and the oxide is represented as Ce_2O_3.

[3]The Oddo—Harkins rule states that elements with an even atomic number (such as carbon) are more common than elements with an odd atomic number (such as nitrogen). Reference: Wikipedia, https://en.wikipedia.org/wiki/Oddo–Harkins_rule See also Oddo (1914) and Harkins (1917).

3.2.2 *Scandium and Yttrium*

The electronic configuration of scandium can be expressed as [Ar] $3d^1$ $4s^2$ leading to oxidation states of 3+ (most common), 2+, and 1+. Its oxide is therefore given as Sc_2O_3. Scandium is considered a rare earth element, but this is firstly because it was discovered together with several other lanthanides. Secondly, scandium resembles yttrium and the rare earth elements more than it resembles aluminium or titanium (Hammond 2015). On the other hand, it also much resembles the ferromagnesian transition elements, although in aqueous systems it behaves more like the REE (see Chap. 1, and McLennan 2012).

The electronic configuration of yttrium can be expressed as [Kr] $4d^1$ $5s^2$ leading to an oxidation state of 3+. Its oxide is given as Y_2O_3

3.3 The Lanthanide Contraction

In the periodic system, the lanthanide group of elements also gives rise to a peculiar phenomenon, called the ***lanthanide contraction.*** This phenomenon is the important and progressive decrease in atomic radii and in radii of ions when going from lower to higher atomic numbers in the lanthanide series. Thus lanthanum has the largest atomic radius, and lutetium has the smallest. In Table 3.3, the ionic radii for the lanthanides are given, and the effect described above can be clearly seen in Fig. 3.2.

In the rows of the periodic system, the valence electrons always shield themselves in an imperfect way from the nuclear charge. This results in an increase in effective nuclear charge, when moving from left to right in a row in the periodic system. The lanthanides show contraction of the atomic and ionic radii, due to the imperfect shielding of the valence f-orbitals. Because there are 14 elements in this series, the effect is more pronounced than other rows of the periodic table. Because the 4f-orbitals are limited in size, the size of the lanthanide ions is defined by their 5s and 5p orbitals (Platt 2012).

The binding energy of an electron to its nucleus is proportional to its mass, so the electrons of the lanthanides are bound more strongly and thus the ionic size is reduced more strongly than would be expected from the increase in nuclear charge and orbital penetration[4] (Platt 2012).

Due to the restricted extension of the 4f orbitals, they cannot overlap with surrounding orbitals of other components. This means that covalent bonding for the lanthanides in their normal oxidation states virtually does not occur. Therefore, in general, the lanthanides are bonded by ionic/electrostatic interactions (Platt 2012).

[4]Orbital penetration is a term that illustrates the proximity of electrons in an orbital to the nucleus. If the penetration for an electron is greater, it experiences less shielding, and therefore a larger effective nuclear charge.

Table 3.3 Ionic radii of the lanthanides in VI-coordination (Shannon 1976)

Element atomic number	La 57	Ce 58	Pr 59	Nd 60	Pm 61	Sm 62	Eu 63	Gd 64	Tb 65	Dy 66	Ho 67	Er 68	Tm 69	Yb 70	Lu 71
2^+ ions						1.22	1.17						1.03	1.02	
3^+ ions	1.032	1.01	0.99	0.983	0.97	0.958	0.947	0.938	0.923	0.912	0.901	0.890	0.880	0.868	0.861
4^+ ions		0.87	0.85												

Values are given in Å

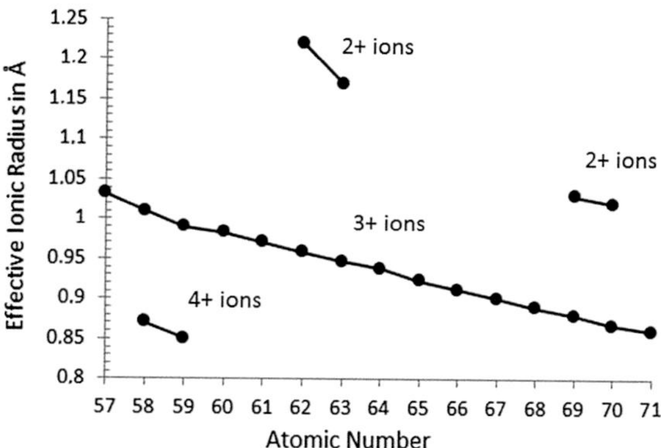

Fig. 3.2 Systematics of effective ionic radii of the lanthanides. The decrease from left to right is called the lanthanide contraction. Ionic radii are according to Table 3.3

Also as a result of the lanthanide contraction, yttrium has an ionic radius comparable to that of the heavier REE species in the holmium-erbium region. If the effective ionic radius (Shannon 1976) of Y^{3+} is plotted (0.90 Å)., it plots in between element 67 (Ho) and 68 (Er). Scandium (effective ionic radius is 0.745 Å), plots outside of the Lanthanide series. As also the outermost electronic arrangement of yttrium is similar to the heavy rare earths, the element behaves chemically like the heavy rare earths. It concentrates during (geo)chemical processes with the heavier REEs, and is difficult to separate from the heavy REEs. Scandium, on the other hand, has a much smaller atomic radius, and the trivalent ionic size is much smaller than that of the heavy rare earths. Therefore, scandium does not occur in rare earth minerals, and in general has a chemical behavior that is significantly different from the other rare earth elements (Gupta and Krishnamurthy 2005).

In sesquioxides (Ln_2O_3, Ln = lanthanide) the coordination number is six (Schweda and Kang 2004).

The lanthanide contraction, however, has also effects for the rest of the transition metals in the lower part of the periodic system. The lanthanide contraction is of sufficient magnitude to cause the elements which follow in the third transition series to have sizes very similar to those of the second row of transition elements. Due to this, for instance hafnium (Hf^{72}) has a 4^+-ionic radius similar to that of zirconium, leading to similar behavior of these elements. Likewise, the elements Nb and Ta and the elements Mo and W have nearly identical sizes. Ruthenium, rhodium and palladium have similar sizes to osmium iridium and platinum. They also have similar chemical properties and they are difficult to separate. The effect of the lanthanide contraction is noticeable up to platinum (Z[5] = 78), after which it no longer noticeable due to the so-called Inert Pair Effect (Encyclopedia Britannica 2015). The inert pair effect describes the preference of post-transition metals to form ions whose oxidation state is 2 less than the group valence.

[5]Z = Atomic Number.

3.4 Radioactivity and Isotopes

Section 3.4 is devoted to the radioactivity and isotopes of the rare earth elements and provides the most important data for the different rare earths. For every element, the tables display the isotopes, their atomic masses, and their abundances.

The information below is taken from Audi et al. (2003). The "Table of Isotopic Masses and Natural Abundances" (www.chem.ualberta.ca/~massspec/atomic_mass_abund.pdf), which shows data from Audi and Wapstra (1993) and Audi and Wapstra (1995), is also used.

3.4.1 *Lanthanum (Element 57)*

Naturally occurring lanthanum (La) has one stable (^{139}La) and one radioactive (^{138}La) isotope. The stable isotope ^{139}La is the most abundant (99.91 % natural abundance). In total, 38 radioisotopes have been found. The most stable ones are ^{138}La with a half-life of 102×10^9 years, ^{137}La with a half-life of 60,000 years and ^{140}La with a half-life of 1.6781 days. The remaining radioactive isotopes have half-lives that are less than a day. The majority of the half-lives of these isotopes are less than 1 min (Table 3.4).

3.4.2 *Cerium (Element 58)*

Naturally occurring cerium (Ce) has 4 stable isotopes: ^{136}Ce, ^{138}Ce, ^{140}Ce, and ^{142}Ce. Of these, ^{140}Ce is the most abundant (88.48 % natural abundance). Other isotopes ^{119}Ce–^{135}Ce, ^{137}Ce, ^{139}Ce, ^{141}Ce, and ^{143}Ce–^{157}Ce are radioactive with half-lives varying from about 100 ns to about 137.6 days (Table 3.5).

3.4.3 *Praseodymium (Element 59)*

Naturally occurring praseodymium (Pr) has only one stable isotope, ^{141}Pr. Thirty-eight radioisotopes are known, of which ^{143}Pr is the most stable with a half-life of 13.57 days and ^{142}Pr with a half-life of 19.12 h. All other isotopes have half-lives shorter than 6 h and for many shorter than 33 s.

Table 3.4 Isotopes of La

Isotope	Atomic mass (u)	Abundance (%)
^{138}La	137.907107	0.090
^{139}La	138.906348	99.910

Table 3.5 Isotopes of Ce

Isotope	Atomic mass (u)	Abundance (%)
^{136}Ce	135.907144	0.185
^{138}Ce	137.905986	0.251
^{140}Ce	139.905434	88.450
^{142}Ce	141.909240	11.114

3.4.4 Neodymium (Element 60)

Naturally occurring neodymium (Nd) has 5 stable isotopes, ^{142}Nd, ^{143}Nd, ^{145}Nd, ^{146}Nd and ^{148}Nd. Of these, ^{142}Nd is the most abundant (27.2 % natural abundance). Two radioisotopes are known, ^{144}Nd and ^{150}Nd. Thirty three radioisotopes of Neodymium are known. The most stable of these are the naturally occurring ^{144}Nd (half-life of 2.29×10^{15} years) and ^{150}Nd (half-life of 7×10^{18} years). The remaining radioactive isotopes have half-lives that are less than 11 days (Table 3.6).

3.4.5 Promethium (Element 61)

This is the only lanthanide that has no stable or even any long-lived isotopes. This is explained below.

Promethium, like other elements with odd atomic numbers $Z^{odd} \geq 9$, should have one or two isotopes, which are stable towards beta decay (further called beta-stable isotopes), with odd atomic numbers M^{odd} having an intermediate value between M^{odd} for beta-stable isotopes for isotopes of neodymium and samarium. However, M^{odd} of beta-stable isotopes of neodymium and samarium follow each other successively: ^{143}Nd, ^{145}Nd, ^{147}Sm, ^{149}Sm. For beta-stable Pm-isotopes, there are no non-occupied numbers left. As a consequence, ^{145}Pm and ^{147}Pm are beta-unstable (Dr. Denis Bykov, TU Delft, RID, pers. comm. 2015). ^{145}Pm transforms by electron-capture decay into ^{145}Nd, and ^{147}Pm by β^--decay transforms into ^{147}Sm. The radioactivity of the synthetic (non-REE) element technetium can be explained in a similar way (Dr. Denis Bykov, TU Delft, RID, pers. comm. 2015).

Table 3.6 Isotopes of Nd

Isotope	Atomic mass (u)	Abundance (%)
^{142}Nd	141.907719	27.2
^{143}Nd	142.909810	12.2
^{144}Nd	143.910083	23.8
^{145}Nd	144.912569	8.3
^{146}Nd	145.913112	17.2
^{148}Nd	147.916889	5.7
^{150}Nd	149.920887	5.6

The explanation given above is also called the *Mattauch Isobar*[6]*Rule*[7] (Mattauch 1934).

The element promethium (Pm^{61}) has 38 radioisotopes. Among them are two relatively stable isotopes, ^{145}Pm, with an atomic weight of 144.9127 and a half-life of 17.7 year, and ^{147}Pm, with an atomic weight of 146.9151, and a half-life of 2.623 year (Wieser and Berglund 2006). Due to the rather short half-life of the two isotopes, promethium does not occur in nature. It can only be synthesized.

Most other isotopes of promethium have a half-life of less than a second. The shortest half- life is for ^{163}Pm, which is approximately 200 ms.

The primary decay products are neodymium and samarium isotopes. Promethium-146 decays to neodymium and samarium (^{146}Nd and ^{146}Sm), the lighter isotopes generally to neodymium via positron decay and electron capture, and the heavier isotopes to samarium via beta decay. Exceptions are ^{130}Pm and ^{131}Pm which give rise to Pr (^{129}Pr and ^{130}Pr respectively) via positron emission. The isotope ^{132}Pm decays to both ^{132}Nd and ^{131}Pr via positron emission, although the majority leads to ^{132}Nd (Oliveira 2011).

3.4.6 Samarium (Element 62)

Samarium has 7 isotopes: ^{144}Sm, ^{147}Sm, ^{148}Sm, ^{149}Sm, ^{150}Sm, ^{152}Sm, and ^{154}Sm. Of these, 4 isotopes are stable, ^{144}Sm, 150Sm, ^{152}Sm, and ^{154}Sm, and three are radioactive isotopes, which, however, are extremely long-lived (^{147}Sm, ^{148}Sm and ^{149}Sm, with half-lifes $>1.0 \times 10^{11}$ years) (Table 3.7).

3.4.7 Europium (Element 63)

Europium has two isotopes ^{151}Eu with an atomic mass of 150.919846 (u) and ^{153}Eu with an atomic mass of 152.921226 (u). The isotope ^{153}Eu is stable, whereas the other was recently found to unstable, but with an extremely long half-life (4.62×10^{18} years; Casali et al. 2014).

---

[6]Isobars refers here to atoms (nuclides) of different chemical elements that have the same number of nucleons (particles in the nucleus, i.e. protons or neutrons). Correspondingly, isobars differ in atomic number (or number of protons) but have the same mass number.

[7]Josef Mattauch (1895–1976) was a German physicist known for his work on establishing the isotopic abundances by mass spectrometry.

Table 3.7 Isotopes of Sm

Isotope	Atomic mass (u)	Abundance (%)
^{144}Sm	143.911995	3.07
^{147}Sm	146.914893	14.99
^{148}Sm	147.914818	11.24
^{149}Sm	148.917180	13.82
^{150}Sm	149.917271	7.38
^{152}Sm	151.919728	26.75
^{154}Sm	153.922205	22.75

3.4.8 Gadolinium (Element 64)

Natural gadolinium consists of 6 stable isotopes, ^{154}Gd, ^{155}Gd, ^{156}Gd, ^{157}Gd, ^{158}Gd and ^{160}Gd, and 1 radioisotope, ^{152}Gd. Its half -life is 10^{12} year (Table 3.8).

3.4.9 Terbium (Element 65)

Naturally occurring terbium has one stable isotope ^{159}Tb with an atomic mass of 158.925343 (u). For terbium, 36 radioisotopes have been characterized. The most stable are ^{158}Tb with a half-life of 180 years, ^{157}Tb with a half-life of 71 years, and ^{160}Tb with a half-life of 72.3 days. All the others have half-lives shorter than 7 days.

3.4.10 Dysprosium (Element 66)

Dysprosium has 7 stable isotopes. There are 29 radioisotopes with the most stable being ^{154}Dy with a half-life of 3.0 million years, ^{159}Dy with a half-life of 144.4 days, and ^{166}Dy with a half-life of 81.6 h. All other radioactive isotopes have half-lives that are less than 10 h (Table 3.9).

Table 3.8 Isotopes of Gd

Isotope	Atomic mass (u)	Abundance (%)
^{152}Gd	151.919788	0.20
^{154}Gd	153.920862	2.18
^{155}Gd	154.922619	14.80
^{156}Gd	155.922120	20.47
^{157}Gd	156.923957	15.65
^{158}Gd	157.924101	24.84
^{160}Gd	159.927051	21.86

Table 3.9 Stable Isotopes of Dy

Isotope	Atomic mass (u)	Abundance (%)
^{156}Dy	155.924278	0.06
^{158}Dy	157.924405	0.10
^{160}Dy	159.925194	2.34
^{161}Dy	160.926930	18.91
^{162}Dy	161.926795	25.51
^{163}Dy	162.928728	24.90
^{164}Dy	163.929171	28.18

3.4.11 Holmium (Element 67)

Naturally occurring holmium has one stable isotope ^{165}Ho with an atomic mass of 164.930319 (u). There are 36 radioisotopes, with the most stable one being ^{163}Ho, with a half-life of 4,570 years. All other radioisotopes have half-lives shorter than 1.2 days.

3.4.12 Erbium (Element 68)

Natural occurring erbium has 6 isotopes. There are 30 radioisotopes known, of which the most stable is ^{169}Er with a half-life of 9.4 days. The other radioisotopes have half-lives shorter than 50 h. The half-life of the majority of the radioisotopes is even shorter than 4 min (Table 3.10).

3.4.13 Thulium (Element 69)

Naturally occurring Thulium has one stable isotope ^{169}Tm with an atomic mass of 168.934211 (u). For thulium, 34 radioisotopes have been identified. Of these the most stable are ^{171}Tm with a half-life of 1.92 years, ^{170}Tm with a half-life of 128.6 days, ^{168}Tm with a half-life of 93.1 days, and ^{167}Tm with a half-life of 9.25 days. All the other radioactive isotopes have half-lives that are less than 64 h, and most of them have half-lives that are less than 2 min.

Table 3.10 Stable isotopes of Er

Isotope	Atomic mass (u)	Abundance (%)
^{162}Er	161.928775	0.14
^{164}Er	163.929197	1.61
^{166}Er	165.930290	33.61
^{167}Er	166.932045	22.93
^{168}Er	167.932368	26.78
^{170}Er	169.935460	14.93

Table 3.11 Stable isotopes of Yb

Isotope	Atomic mass (u)	Abundance (%)
^{168}Yb	167.933894	0.13
^{170}Yb	169.934759	3.04
^{171}Yb	170.936322	14.28
^{172}Yb	171.936378	21.83
^{173}Yb	172.938207	16.13
^{174}Yb	173.938858	31.83
^{176}Yb	175.942568	12.76

Table 3.12 The stable and the virtually stable isotopes of Lu

Isotope	Atomic mass (u)	Abundance (%)
^{175}Lu	174.940768	97.41
^{176}Lu	175.942682	2.59

3.4.14 Ytterbium (Element 70)

Ytterbium has 7 stable isotopes. There are 27 radioisotopes known. The most stable are ^{169}Yb with a half-life of 32.026 days, ^{175}Yb with a half-life of 4.185 days, and ^{166}Yb with a half-life of 56.7 h. All of the remaining radioactive isotopes have shorter half lives.

3.4.15 Lutetium (Element 71)

Natural lutetium consists of two isotopes: ^{175}Lu and ^{176}Lu (Tables 3.11 and 3.12). The latter is actually radioactive, but has an extremely long half-life of 3.78×10^{10} years. In total, 34 radioisotopes have been identified, of which, besides ^{176}Lu, the most stable are ^{174}Lu with a half-life of 3.31 years, and ^{173}Lu with a half-life of 1.37 years. All of the remaining radioactive isotopes have half-lives that are less than 9 days. Many half-lives are even less than half an hour.

3.5 Magnetism

For most of the lanthanide elements, their Curie-temperatures,[8] as far as they are known, are usually quite low, which results in the metals at room temperature generally displaying paramagnetism. An exception is gadolinium (Gd), which has a

[8]Ferromagnetism is the magnetism commonly observed in iron, nickel and cobalt. The following types of magnetism are also recognized: ferrimagnetism, antiferromagnetism, and paramagnetism.

Table 3.13 Curie temperatures of the lanthanides. Data from Jensen and Mackintosh (1991)

	T_c (Curie-point) (K)	Type of magnetism
La	Not available	paramagnetic
Ce	Not available	paramagnetic
Pr	Not available	paramagnetic
Nd	Not available	paramagnetic
Pm	Not known	Not Known
Sm	Not available	paramagnetic
Eu	Not available	paramagnetic
Gd	293 K	ferromagnetic
Tb	220 K	paramagnetic
Dy	89 K	paramagnetic
Ho	20 K	paramagnetic
Er	20 K	paramagnetic
Tm	32 K	paramagnetic
Yb	Not available	paramagnetic
Lu	Not available	paramagnetic
Y	Not available	paramagnetic

Curie-temperature of 292 K (18.9 °C). It is ferromagnetic. The only other lanthanide with a (rather) high Curie-temperature is terbium, which has a Curie temperature of 222 K (−51.1 °C). The other known values are below 87 K (−186 °C). In Table 3.13, the known Curie-temperatures of the lanthanides are given.

3.6 Chemical Behavior

The rare earth elements are very electropositive, and, as a consequence, they generally form ionic compounds. Mineralogically, the REEs therefore form oxides, halides, carbonates, phosphates and silicates, borates or arsenates, but not sulphides. (Henderson 1996). Their oxidation states are given in Table 3.2.

(Footnote 8 continued)

The first two occur only below the Curie-temperature, the last one only above. On heating. a ferromagnetic, ferrimagnetic or antiferromagnetic material will convert to a paramagnetic material at the Curie-temperature. A paramagnetic material is only magnetic when placed in a magnetic field. Antiferromagnetic substances are actually not magnetic at all. This is because of a special organization of electron spins with opposite sign, which carry the magnetic moments. The magnetic moments of these electron spins cancel each other out, contrary to ferrimagnetism, where they do not, resulting in the case of ferrimagnetism as an observable magnetism.

3.6.1 Air and Oxygen

At room temperature, not all the rare earth metals are affected by air in the same way. The light REEs tend to tarnish very quickly, especially europium, but also lanthanum and neodymium. When the air is moist, and at elevated temperatures, oxidation proceeds even more rapidly. Oxidation increases by a factor of ten when the relative humidity is increased from 1 to 75 % (Gupta and Krishnamurthy 2005). Rare earth oxides do not all have the same structure, and therefore some oxide coatings on fresh metal surfaces in contact with water vapor spall, exposing fresh surfaces, whereas others form a persistent tight layer, which prevents further oxidation. Rare earth oxides have a very large negative free energy of formation, even belonging among the most negative in the whole periodic system. As a consequence, the rare earth oxides are among the most stable in the periodic system (Gupta and Krishnamurthy 2005).

3.6.2 Nitrogen

The REEs also have a strong affinity with nitrogen. The mononitrides of the rare earths are very stable and comparable in stability to those of titanium or zirconium (Gupta and Krishnamurthy 2005). The reaction with nitrogen is, however, slow. High temperatures are needed for a reaction to proceed at a visible rate. The nitrides also form a stable layer on the surface, blocking any further nitridation. Solid solutions with nitrogen are also possible (Gupta and Krishnamurthy 2005).

3.6.3 Hydrogen

Rare earth metals readily form hydrides at temperatures of 400–600 °C. Several of them decompose and degas relatively easily (Gupta and Krishnamurthy 2005).

3.6.4 Carbon

All rare earths easily form dicarbides ($REEC_2$). Several of them (La–Sm and Gd–Ho) also form sesquicarbides (REE_2C_3). Solid solubility of carbon in rare earths also occurs easily. REE-carbides also form solid solutions with nitrogen and oxygen (Gupta and Krishnamurthy 2005).

3.6.5 *Silicon*

Silicon (Si) forms silicides and solid solutions with all REE. Silicides are generally disilicides, $REESi_2$ (Gupta and Krishnamurthy 2005).

3.6.6 *Refractory Metals*

Niobium, molybdenum tantalum and tungsten are refractory metals, i.e. the ones that show resistance against attack by liquid RE metals. These metals are usually listed in the literature in order of decreasing solubility in the liquid rare metals at high temperatures. Tungsten is the least soluble. However, tungsten is rather brittle and has poor mechanical properties compared to tantalum, which is the second best with respect to solubility. Tantalum containers are therefore used in purification of liquid rare element metals (Gupta and Krishnamurthy 2005).

3.6.7 *Acids and Bases*

The REE readily dissolve in dilute mineral acids. Concentrated sulphuric acid has a somewhat smaller effect on rare element metals. The metals resist attach by HF, because a thin layer of $REEF_3$ forms on the metal, preventing further dissolution (Gupta and Krishnamurthy 2005).

3.6.8 *Water*

The reactions of the rare earths with water varies, depending on the metal. LREE react with water slowly at room temperature and more strongly at higher temperatures. HREE react very slowly with water (Gupta and Krishnamurthy 2005).

In aqueous systems, the typically trivalent rare earths show strong ionic character. Ce (IV) is the only tetrapositive rare earth species that is stable in aqueous solution as well as in solids. The trivalent rare earths form salts with a large number of anions. There is a wide variability in solubility of such salts. Rare earths containing thermally unstable ions like OH-, CO_3^{2-}, or $C_2O_4^{2-}$, convert, when heated first, into the basic derivative, and finally into the oxide (Gupta and Krishnamurthy 2005). Chlorides, bromides, nitrates, bromates, and perchlorate salts of rare earths are all water soluble. When crystallizing as a result of evaporation, they all form hydrated crystalline salts (Gupta and Krishnamurthy 2005).

Fig. 3.3 Frequency and wavelength range

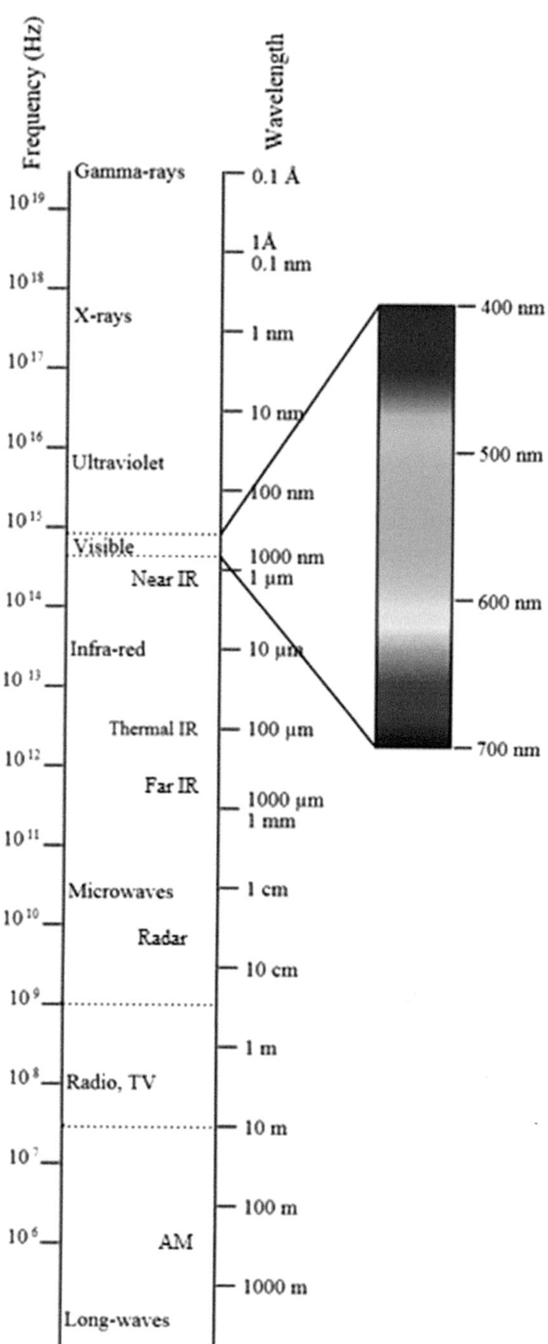

Table 3.14 Summary of the main visible emissions of lanthanide ions, and typical intensities of their visible emission (Andres and Chauvin 2012)

Trivalent lanthanide ion	Colour of visible luminescence	Intensity of the emission
Eu	Red or red-orange	Strong
Tb	Green	Strong
Dy	Yellow	Medium
Sm	Orange-red	Medium
Pr	Red	Weak
Er	Green	Weak
Ho	Red	Weak
Tm	Blue	Weak

3.7 Luminescence

The emission of light by a body can result from heat: for instance iron may glow red-yellow when heated to high temperatures. This is called incandescence. Luminescence is the emission of light by a substance which is not the result of heating. It is a type of cold-body radiation.

The lanthanides display luminescence in the spectral range from ultraviolet (UV) to near infrared (NIR). The spectral ranges are displayed in Fig. 3.3. The near infrared is a subsection of the infrared spectrum, ranging in wavelength from 0.75–1.4 μm.

Many lanthanide ***ions*** have unique spectral characteristics in the visible region of the electromagnetic spectrum, which results in their having distinct luminescent colors (Andres and Chauvin 2012). Many applications are based on these characteristic emissions for color reproduction and lighting. Among them are phosphors and LEDs. Another intensively studied application is energy conversion. The lanthanide ion converts one type of excitation energy (UV or IR) into visible energy that can be easily be absorbed by a photosensitive instrument, for instance a solar cell (Andres and Chauvin 2012).

With respect to luminescence, the most used lanthanides are europium and terbium. They have red (europium) and green (terbium) emissions. They are applied in additive color synthesis where all colors are reproduced by combinations of red, green and blue. Dysprosium (yellow) and samarium (orange) are also interesting for their luminescence, but the emissions are weaker than for europium and terbium. Other REEs exhibiting luminescence are thulium (red, blue, NIR), and praseodymium, but they have not been fully investigated (Andres and Chauvin 2012).

In the NIR part of the spectrum, the best lanthanides are ytterbium, erbium and neodymium. Holmium has been less investigated, but shows an emission spectrum with a peak in the red and a peak in the IR.

Finally, divalent europium exhibits luminescence in the blue part of the visible spectrum (Andres and Chauvin 2012).

The colors of the emissions and their intensities are summarized in Table 3.14.

References

Andres J, Chauvin AS (2012) Lanthanides: luminescence applications. In: Atwood DA (ed) The rare earth elements—fundamentals and applications. Wiley, Chichester, pp 135–152

Audi G, Bersillon O, Blachot J, Wapstra AH (2003) The NUBASE evaluation of nuclear and decay properties. Nucl Phys A 729:3–128

Audi G, Wapstra AH (1993) The 1993 atomic mass evaluation. Nucl Phys A 565:1–65

Audi G, Wapstra AH (1995) The 1995 update to the atomic mass evaluation Nucl. Phys A 595:409–480

Beaudry BJ, Geschneider KA Jr (1978) Preparation and basic properties of the rare earth metals. In: Gschneidner KA Jr, Eyring L (eds) Handbook on the physics and chemistry of rare earths, chapter 2. North-Holland Publishing Company, pp. 174–232

Binnemans K, Jones PT, Van Acker K, Blanpain B, Mishra B, Apelian D (2013) Rare-Earth economics: the balance problem. JOM 65(7):846–848

Casali N, Nagorny SS, Orio F, Pattavina L, Beeman JW, Bellini F, Cardani L, Dafinei I, Di Domizio S, Di Vacri ML, Gironi L, Kosmyna MB, Nazarenko BP, Nisi S, Pessina G, Piperno G, Pirro S, Rusconi C, Shekhovtsov AN, Tomei C, Vignati M (2014) Discovery of the 151Eu α decay. J Phys G: Nucl Part Phys 41, 1–8

Choppin GR, Rizkalla EN (1994) Solution chemistry of actinides and lanthanides. In: Gschneider KA, Eyrting L, Choppin GR, Lander GR (eds) Handbook of the physics and chemistry of rare earths, vol 18, chapter 128, pp 559–589

Encyclopedia Britannica (2015) http://www.britannica.com/science/nitrogen-group-element (Accessed Sep 2015)

Gupta CK, Krishnamurthy N (2005) Extractive metallurgy of rare earths. CRC Press, Boca Raton, chapter 1.5, pp 22–25

Haire RG, Eyrting L (1994) Comparisons of the binary oxides. In: Gschneider KA, Eyrting L, Choppin GR, Lander GR (eds) Handbook of the physics and chemistry of rare earths, vol 18, chapter 125, pp 413–505

Hammond CR (2015) The Elements. In: Handbook of Chemistry and Physics, 96th edition, p 4–32

Harkins WD (1917) The evolution of the elements and the stability of complex atoms. J Am Chem Soc 39(5):856–879

Henderson P (1996) The rare earth elements: introduction and review. In: Rare earth minerals, chemistry, origin and ore deposits, The Mineralogical Society Series, vol 7. Chapman and Hall, London, chapter 1, pp 1–19

Jensen J, Mackintosh A (1991) Rare Earth magnetism—structures and excitations. Clarendon Press, Oxford 413 p

Jensen WB (2007) The origin of the s, p, d, f orbital labels. J Chem Educ 84:757–758

Madelung E (1943) Die Mathematischen Hilfsmittel des Physikers (Mathematical Tools for the Physicist), 3d improved and extended edition. Dover Publications, New York, p 359

Mattauch J (1934) Zur Systematik der Isotopen. Z Phys 91(5–6):361–371

McLennan SM (2012) Geology, geochemistry, and natural abundances of the rare earth elements. In: Atwood DA (ed) The rare earth elements—fundamentals and applications. Wiley, New York, pp 1–19

Oddo G (1914) Die Molekularstruktur der radioaktiven Atome. Zeitschrift für Anorganische Chemie 87:253–268

Oliveira P (2011) The Elements. Pediapress, p 683

Platt AWG (2012) Group trends. In: Atwood DA (ed) The rare earth elements—fundamentals and applications. Wiley, Chichester, pp 44–53

Shannon RD (1976) Revised effective ionic radii and systematic studies of interatomic distances in halides and chalcogenides. Acta Cryst, A, 32:751–767

Schweda E, Kang Z (2004) Structural features of rare earth oxides. In: Adachi G, Imanaka N, Kang ZC (eds) Binary rare earth oxides. Kluwer Academic Publishers, Dordrecht, chapter 3, pp 57–93

The Spectrum of Riemannium https://thespectrumofriemannium.wordpress.com/?s=Madelung (Accessed Aug 2014)

Wieser ME, Berglund M (2006) Atomic weights of the elements 2005. IUPAC Technical Report. Pure Appl Chem 78(11):2051–2066

Chapter 4
Mineral Processing and Extractive Metallurgy of the Rare Earths

Abstract This chapter is about mineral processing of the rare earths (making the mined ore into a concentrate of the valuable minerals), and extractive metallurgy of the rare earths (how to get the metals out of the concentrate). The mineral processing of three well-known exploited ore deposits is discussed in more detail.

4.1 Introduction

In Chap. 2, the most important occurrences of the rare earths were described. This chapter will deal with what happens when the rare earths have been mined. It answers the question: how are the ores processed?

Mineral processing (sometimes called mineral dressing or ore dressing) is the preparation of ore from the stage that the material is brought out of the mine (also called "run of mine" or "as-mined") through the production of a mineral concentrate, from which the desired metals can be extracted.

To be able to extract a valuable metal from an ore, the ore as it comes from the mine (usually in fairly large lumps), must be broken down to smaller sizes, and then it must be ground to a finer grain size that is suitable for the separation techniques employed thereafter. The separation step is necessary to separate the valuable minerals from the worthless bulk. Most of the ore is usually worthless. The worthless material is collectively termed "gangue". Gangue must be distinguished from overburden, which is waste material overlying an ore or mineral body. These latter materials are just displaced, not processed (Wills and Napier-Munn 2006). Ores commonly contain just a few percent of the valuable material, although very rich deposits can sometimes contain 5–15 % ore (see Chap. 2). These rich ores, however, may again be located in veins, which cannot be selectively mined, and, as a consequence, a part of country[1] rocks has to be mined too, in order to have an efficient operation.

[1]Country rocks are the rocks around the ore deposit. They consist of non-valuable material.

J.H.L. Voncken, *The Rare Earth Elements*, SpringerBriefs in Earth Sciences,
DOI 10.1007/978-3-319-26809-5_4

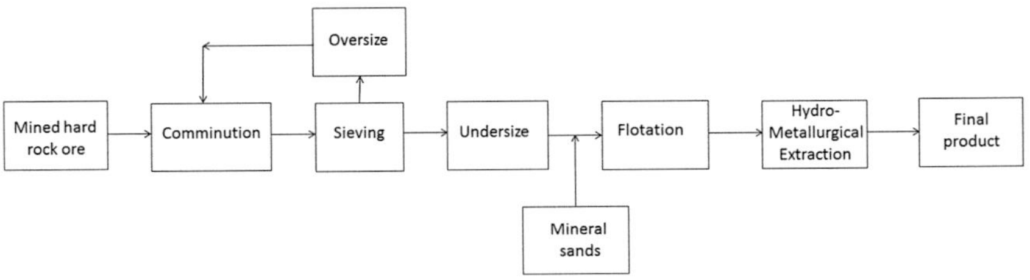

Fig. 4.1 Basic flow sheet of ore processing. Instead of hydrometallurgical processing, pyrometallurgical processing, (involving heat) may be applied (c.f. iron ore processed with a blast furnace)

4.2 Mineral Processing and Extraction of Rare Earths

In this section, it is explained how hard-rock deposits of rare earths are processed. Ores that consist of hard rocks, have to undergo several procedures before the concentration of the valuable material and the extraction of the valuable material from this concentrate can begin. Figure 4.1 gives a basic flow sheet of the process.

4.2.1 *Hard Rock Deposits*

An intimate knowledge of the mineralogical assembly of the ore is necessary if efficient processing is to take place. Not only the nature of the valuable minerals and gangue must be known, but also knowledge of the texture of the ore is required. The texture refers to the size, dissemination, association and shape of the minerals within the ore. The processing of the ore should always be considered in relation to the mineralogy (Wills and Napier-Munn 2006).

4.2.1.1 Liberation

Liberation of the valuable minerals from the gangue is achieved by comminution.[2] The particles resulting from this process step should be of such as size as to yield relatively clean particles of either mineral or gangue. Grinding is usually the most expensive operation in mineral processing. It may account for up to 50 % of the energy used by the concentrator plant (Wills and Napier-Munn 2006). The relative degree of liberation resulting from a grinding operation is pictured in Fig. 4.2.

[2]Comminution means crushing or pulverization.

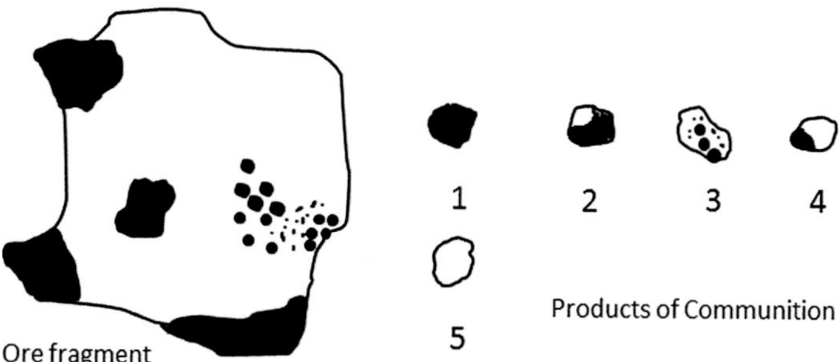

Fig. 4.2 Cross sections of ore particles, before (*left*) and after comminution (*right*). *Black* areas represent valuable mineral

To the left, a large fragment from run of the mine[3] is shown and to the right four fragments resulting from grinding. Black areas represent valuable mineral. The two regions with many small particles of ore and the black spots, within the largest fragment 3, represent highly intergrown areas of ore and gangue. Comminution produces a wide range of particles, varying from completely liberated ore, to completely liberated gangue, and intermediate particles as shown in Fig. 4.2. A particle like particle 1 could be used for metal extraction, particle 3 and 5 would be considered waste (tailings), and particles 2 and 4 might be reground. Regrinding will depend on the relative amounts of different types of particles, and the grade of the ore. If, for instance, the ore would look completely like comminution product 3, one would definitely regrind. If this only a minor proportion of the ore looks like this, the value of the material will play a determining role.

After comminution, sieving operations separate large-, mainly poorly-ground particles from smaller wellground particles. Particles that are too large are reground.

4.2.1.2 Mineral Separation (Flotation)

After comminution, mineral separation will have to take place, to separate the valuable particles from the non-valuable gangue particles. As the ore minerals of the rare earths are hardly magnetic, useful separation techniques would be gravity separation or flotation. For bastnaesite, which does not have an extremely high density, flotation is applied. The principle of flotation is depicted in Fig. 4.3.

Bastnaesite from Mountain Pass ore (hard-rock ore) is processed as follows. After comminution, the ground ore is subjected to hot-froth flotation. However, gangue minerals like calcite, barite or celestite may cause problems in flotation, because they have flotation properties similar to bastnaesite. For this, special treatment has been devised, consisting of six different (steam) conditioning steps

[3]Run of mine is the raw product that comes from the mine. This may be material like the ore fragment on the left in Fig. 4.2.

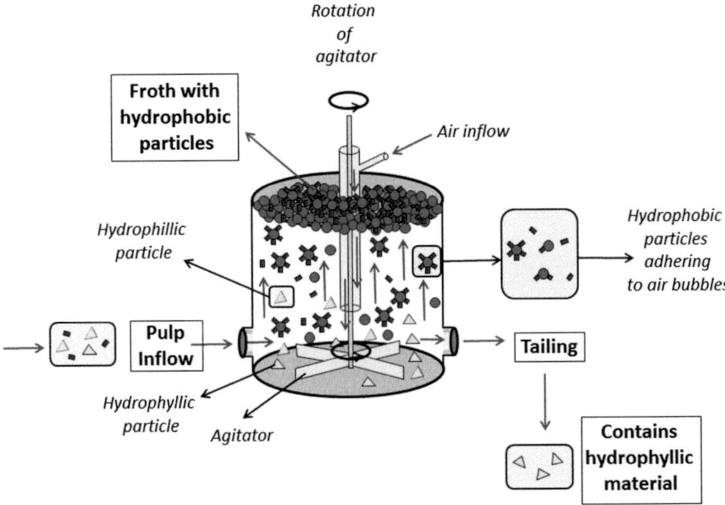

Fig. 4.3 The principle of froth flotation. Redrawn after Encyclopedia Brittanica (2015a)

(Gupta and Krishnamurthy 2005). This yields a slurry containing 30–35 % solids, which is then fed into the flotation process.

In flotation, first a slurry of desired density is formed by adding mineral powder to water. Next, chemicals are added to the diluted slurry in a tank, through which air bubbles are pumped. The slurry is agitated by a stirrer. Due to the added chemicals, the desired mineral(s) become hydrophobic, and adhere to the bubbles, whereas the unwanted minerals become hydrophilic. Also other specific chemicals may be used, which make certain minerals intentionally hydrophyllic, thus preventing them from floating. Such chemicals are termed depressants. The bubbles with the hydrophobic mineral grains rise to the surface, where they form foam, which either overflows the cell or is skimmed off. If more than one mineral must be separated, the process has to be repeated several times. Flotation plants normally perform these separations in several steps, namely rougher, scavenger and cleaner. These comprise the first cells in the bank (rougher), the second set of cells (scavengers), and finally the cleaners, which have diluted pulp, and are designed to take out the last of the valuable mineral (Wills and Napier-Munn 2006).

4.2.1.3 Mineral Processing at Mountain Pass, Bayan Obo, and Mount Weld

At Mountain Pass, a four-stage cleaning takes place, after which the mineral concentrate is thickened, filtered, and dried. The final dried concentrate of Mountain Pass contains 60 % REO (rare earth oxide). The overall recovery is 65–70 %, which means that 65–70 % of the valuable material in the ore ends up in the concentrate.

The quality of the concentrate is expressed as the grade of the concentrate, which gives the pureness of the concentrate (Gupta and Krishnamurthy 2005).

Bastnaesite from Bayan Obo is strongly intergrown with other minerals. These are magnetite (Fe_3O_4), fluorite (CaF_2), hematite (Fe_2O_3), and various niobium oxide minerals (Zhang et al. 2002) and are not waste, but are also valuable. They are recovered as by-products.

The ore is ground to a state where 90 % is smaller than 74 μm. pH is regulated and depressants are applied for the gangue.[4] A collector is applied to amass the valuable minerals in the foam. Depressed iron oxide minerals and silicate minerals remain at the bottom of the flotation cells and are removed for iron beneficiation and niobium recovery.

After thickening and desliming,[5] selective rare-earth flotation is carried out. In this phase, calcite and fluorite and barite are depressed. The recovery of rare earths as concentrate in this stage is approximately 80 % (Gupta and Krishnamurthy 2005). After this phase, final cleaning is carried out by high-gradient magnetic separation, and two fractions emerge: a 68% REO concentrate and a secondary monazite concentrate, containing 36 % REO. The recoveries are 25 and 36 % respectively. Overall recovery of rare earths from the ore is 61% (Gupta and Krishnamurthy 2005).

At Mount Weld, Australia, physical concentration of the REE minerals is performed as follows: the run of mine material is subjected to crushing, grinding and *flotation* to produce a concentrate with a grade of 40 % REO. The obtained *flotation* concentrate is sent to a thickener, where the resulting pulp is dewatered by applying pressure filtration.

4.2.2 *Placer Deposits*

Heavy mineral sands (placers) in wet environments are mined by dredging techniques, applying bucket line and suction dredges (relatively deep water), or bucket wheel units (relatively shallow water). In dry environments, open pit excavation methods are used. Drilling and blasting are in general not required, but may be applied when the sand is strongly cemented. Mineral sands are processed further as-mined and are ***not*** subjected to any comminution (Gupta and Krishnamurthy 2005).

Heavy mineral sands may differ in composition, depending on the location, and only generalized descriptions of the processing can be given. Monazite and

---

[4]Gangue: this is the commercially worthless material that surrounds, or is closely mixed with, a desired mineral in an ore deposit. For instance in so-called Banded Iron Ore, the iron-ore minerals are magnetite (Fe_3O_4) and hematite (Fe_2O_3), and the gangue is quartz (SiO2). Gangue should not be confused with overburden, which is the material that covers a deposit.

[5]Desliming is the removal of the finest fraction, which often may cause difficulties in later processing stages.

xenotime from heavy minerals sands are separated from the other commonly occurring heavy minerals like ilmenite, zircon, rutile, garnet, magnetite and quartz. Techniques used for heavy mineral-sand separation utilize slight differences in specific gravity, magnetic properties, or electrical properties. The specific gravity of monazite is the highest. Ilmenite, garnet, xenotime, and monazite, in decreasing order of magnetisability, behave as magnetic minerals. In electrostatic separation, ilmenite and rutile behave in some instances as conducting minerals, in other instances as non-conducting minerals. Xenotime is more strongly magnetic than monazite and concentrates with ilmenite in magnetic separation. As xenotime is a poor conductor, electrostatic separation is applied to separate it from ilmenite. Leucoxene,[6] when present, may cause problems in the separation of monazite from ilmenite. In that case, separation may be enhanced by roasting under reducing conditions at 600 °C. This converts free hematite in leucoxene into magnetite, which eases the concentration. Flotation is occasionally used (Gupta and Krishnamurthy 2005).

4.2.3 *Chemical Treatment to Decompose the Mineral Concentrates*

In this subsection, the chemical treatment of the rare earths is addressed per ore mineral. The text is largely a summary of chapter three from *Extractive Metallurgy of the Rare Earths* by Gupta and Krishnamurthy (2005).

4.2.3.1 Monazite

Monazite is (Ce, La, Th)PO_4., which is the ideal formula. To extract the rare earths and to remove thorium, a variety of methods are applied after chemically attacking the mineral with either sulphuric acid, or sodium hydroxide.

Acid Treatment

The sulphuric acid method has been used most extensively in the USA. With this method, depending on the ratio of acid to ore, temperature and concentration, the rare earths can either be solubilized, or the thorium can be extracted, or both can be brought into solution (Gupta and Krishnamurthy 2005).

---

[6]Leucoxene: an alteration product and mixture of Fe-Ti oxides, including titanite, perovskite, titanian magnetite, but especially ilmenite. Most leucoxene is actually largely anatase, TiO_2, or rutile, TiO_2 (Mindat.org 2015). Fe occurs usually as hematite.

The process of rare earth recovery is based on rare-earth double-salt precipitation. However, yttrium and the heavy rare earths go with thorium. The rare earths are recoverable from the thorium fraction during the solvent extraction[7] step used for the purification of uranium and thorium. Solvent extraction with TBP (tributyl phosphate[8]), from an aqueous 8 N nitric acid solution of thorium and mixed rare earths, enables the recovery of thorium, uranium, cerium and cerium free rare earths (Gupta and Krishnamurthy 2005). Other significant processes involve precipitation of thorium pyrophosphate,[9] or precipitation as basic salts from the leach liquor. After that comes recovery of the rare earths from solution as double sulphates, fluorides, or hydroxides, and also selective solubilisation of thorium itself in the ore treatment stage. The sulphuric acid route does yield impure products, but it is not used anymore (Gupta and Krishnamurthy 2005).

Alkali Treatment

In this process, caustic soda (NaOH) is used. The phosphate part of the ore is recovered as trisodium phosphate. This product is marketable, which is already a major advantage over the acid process, and this process has been very favorable for commercial use. Fine-ground monazite is treated with a 60–70 % NaOH solution at 140–150 °C. The mixed rare earth thorium hydroxide cake resulting from this process is then treated for recovery of thorium and rare earths. A variety of methods is used to accomplish this (Gupta and Krishnamurthy 2005). An effective process for removing thorium completely and in a very pure state is solvent extraction with higher amines. This should preferably be carried out in a sulphate solution. However, leaching of the rare earths from the hydroxide cake with hydrochloric acid is also very effective. The rare earths are recovered from the leachate by solvents extraction.

Several varieties of the alkali treatment method exist, among them the IRE-process (Indian Rare Earths Ltd). These are described in detail in Gupta and Krishnamurthy (2005).

4.2.3.2 Bastnaesite

After physical processing of bastnaesite (ideal formula (Ce, La, Y)CO_3F), a 60 % concentrate is obtained. This can be upgraded by leaching with hydrochloric acid to 70 % REO, while eliminating calcium and strontium carbonates. After calcination, the percentage of REO can be upgraded to 85–90 %. This occurs because calcination removes the carbon dioxide (Gupta and Krishnamurthy 2005).

---

[7]Solvent extraction is a separation method in which a substance is separated from one or more others using a solvent. Fundamental to the method is the variation of solubilities of different compounds in different substances.

[8]Tributyl phosphate is an organophosphorus compound with the chemical formula $(C_4H_9)_3PO_4$ (Encyclopedia Britannica 2015b).

[9]A pyrophosphate is a phosphorus oxyanion (Wikipedia 2015a).

U.S. Bureau of Mines Processes
The decomposition of bastnaesite has been intensively investigated. At the U.S. Bureau of Mines at the end of the 1950s, two processes were developed. One involved nitric-acid digestion and solvent extraction, the other involved sulphuric acid and recovery of the rare earths as water soluble sulphates (Gupta and Krishnamurthy 2005).

Molycorp Process-1
Molycorp developed another process, in which the concentrate is roasted in air at 620 °C to drive off CO_2, and oxidize cerium in to a tetravalent state. The resulting material is treated with 30% HCL to dissolve the non-cerium rare earths. This results in a cerium concentrate containing 65–70 % REO with 55–60 % CeO_2. The solution is further processed for the recovery of other rare earths, among them europium. The residue also contains RE-fluorides. The fluorides are decomposed by treatment with caustic soda. The rare earth hydroxides are then leached with HCl (Gupta and Krishnamurthy 2005).

Molycorp Process-2
Another process, developed at Molycorp, consists of the treatment of bastnaesite with HCl, yielding rare earth chlorides. These are subsequently treated with NaOH to convert them onto RE-hydroxides. After separation, are dissolved again in HCl, yielding rare earth chlorides. In the next step, the rare earth fluoride in the solid residue is converted to rare earth hydroxide using NaOH. Next follows neutralisation and purification. This involves hydrochloric acid, yielding a solution with a pH of about 3. By addition of hydroxide, iron precipitates as iron hydroxide. Sulphuric acid is used to precipitate lead sulphate. Then barium chloride is added to precipitate excess sulphate and to act as a carrier for the removal of any thorium daughter products present in the ore. At this pH, thorium hydroxide is insoluble and can be removed. Filtration finally leads to a clear solution of rare earth chlorides. This solution is then either concentrated by evaporation or by evaporation made into a solid form (Gupta and Krishnamurthy 2005).

Processing in China
In China, the rare earths are processed by heating with 98 % sulphuric acid at 500 ° C in a rotary kiln. This results in the release of carbon dioxide and hydrofluoric acid. The rare earths are converted to their sulphates. They are precipitated as a double sodium sulphate by leaching with water and adding sodium chloride. Subsequently, the sulphates are converted to hydroxides by digestion in a strong caustic solution. The separated hydroxides are then dissolved in HCl. After this, separation and purification of the rare earths is done through solvent extraction (Gupta and Krishnamurthy 2005).

Thorium Ltd Process
Bastnaesite is treated with caustic soda to convert the fluoride component into hydroxide. The hydroxide is treated with HCl, which yields a rare earth chloride solution. From this, $RECl_3 \cdot 6H_2O$ is produced, or the solution is processed further for the recovery of the individual rare earths (Gupta and Krishnamurthy 2005).

4.2.3.3 Chlorination

Goldschmidt process

The firm Goldschmidt AG from Germany developed in 1967 a high-temperature chlorination process to directly obtain anhydrous $RECl_3$. From this, metal can be produced directly. The process is suitable for a variety of ore minerals, such as monazite, xenotime, allanite, euxenite, fergusonite and gadolinite.

The ground ore concentrate is thoroughly mixed with a binder and some water. The relative dry brew is compacted into pellets in a briquetting machine, and then passed through a band drier. Eventual fines are screened off, and returned to the mixer. Then the pellets, accompanied by carbon, are fed into the chlorination furnace, where they are converted into chlorides by gaseous chlorine. The chlorides are separated according to their volatility. The rare-earth chlorides are periodically tapped from the bottom of the furnace. Alkali and alkaline earth also collect in the melt chamber. All other reaction products are carried away by the off gases (Gupta and Krishnamurthy 2005).

Baotau Concentrates

Baotau is a city in the REE-rich area of Inner Mongolia in China. The process developed here is similar to the process mentioned above. Ore minerals are mainly bastnaesite and monazite (Gupta and Krishnamurthy 2005).

4.2.3.4 Xenotime

Xenotime (ideal formula YPO_4) is processed differently by using concentrated sulphuric acid. The double sulphate precipitation is not possible, because yttrium and the heavy rare earths are very soluble. The rare earth phosphates are converted into water soluble sulphates by leaching xenotime concentrates in concentrated H_2SO_4 at 250–300 °C for one—two hours. The sulphate solutions are directly used for separation.

In an alternative process, the fine-ground xenotime is treated by fusing it with molten caustic soda at 400 °C, or by mixing it with sodium carbonate and roasting at 900 °C for several hours. The hydroxide residue is dissolved in hydrochloric acid and filtered to remove impurities such as silica, cassiterite etc. The rare earths are precipitated as oxalates by adding oxalic acid. The oxalates are oxidized to rare earth oxides (Gupta and Krishnamurthy 2005).

4.2.3.5 Gadolinite

For processing of gadolinite (ideal formula (Ce, La, Nd, $Y)_2FeBe_2Si_2O_{10}$), the ground ore is leached with nitric or hydrochloric acid. The rare earths are precipitated as oxalates by adding oxalic acid, which makes possible the separation of associated iron and beryllium. Concentrated sulphuric acid or mixtures of sodium hydroxide and sodium peroxide have also been applied to attack the ore (Gupta and Krishnamurthy 2005).

4.2.3.6 Euxenite

Euxenite (ideal formula (Y, Ca, Ce)(Nb, $Ta)_2O_6$), is treated by reductive chlorination, which is followed by distillation of the chlorides to separate the rare earths, titanium, niobium and tantalum. Solvent extraction can be used to remove tantalum and niobium chlorides and the rare earths from the chlorination residue (Gupta and Krishnamurthy 2005).

The ore mineral can also be treated by fusing it with ammonium bisulphate or caustic soda, or by digestion with hydrofluoric acid or sulphuric acid, or mixtures of the two (Gupta and Krishnamurthy 2005).

Also digestion in hot sodium hydroxide has been tried. The hydroxides formed are dissolved by dilute hydrochloric acid. Soda ash is added to precipitate a complex carbonate precipitate. This in turn is leached with dilute sulphuric acid, to selectively solubilize the rare earths. The rare earths are then subsequently precipitated as oxalate by adding oxalic acid (Gupta and Krishnamurthy 2005).

4.2.3.7 Loparite, Pyrochlore, Fergusonite and Samarskite

Loparite may be decomposed by chlorination. Pyrochlore decomposition involves use of hydrofluoric acid, followed by solvent extraction of Nb by MIBK (methylisobutylketone). There is also another process, which involves digestion with hot sulphuric acid and precipitation of niobium and rare earths by gradual reduction of acid concentrate and temperature.

Extraction of the rare earths from samarskite and fergusonite is carried out by hot caustic digestion and subsequently acid dissolution. This process is similar to that described for euxenite.

4.2.3.8 Apatite

Apatite may contain significant amounts of rare earths. Examples are apatites from Palabora, RSA (Gupta and Krishnamurthy 2005), Kovdor, Russia (Lapin and Lyagushkin 2014), and Kagankunde, Malawi (Wall and Mariano 1996).

In Palabora, the apatite is decomposed as follows. Apatite is dissolved in a mixture of sulphuric acid and recycled dilute phosphoric acid from the wet process phosphoric acid plant. Most of the rare earths, (70–85 %) contained in the feed, end up in largely in phosphogypsum, which is formed in the dissolution reaction. The rest of the rare earths dissolve in the crude dilute phosphoric acid and eventually precipitate in sludge that forms when the acid is concentrated to commercial-grade phosphoric acid (Gupta and Krishnamurthy 2005).

This sludge contains 2–8 % rare earths. As large amounts of this sludge are produced, they are a valuable source of rare earths. The sludge is washed and leached with dilute nitric acid, to which calcium nitrate has been added. From the leach liquor, the rare earths are recovered by solvent extraction and finally precipitated as rare earth oxalates. The oxalates are calcined in a rotary furnace to yield mixed rare earth oxides of 89–94 % purity.

4.3 Separation Processes

Several processes exist for separation of the naturally occurring rare earths from one another. They all utilize the small differences in basicity resulting from the decrease in ionic radius from lanthanum to lutetium. The basicity differences cause several effects:

- differences in solubility of salts
- differences in the hydrolysis of ions and
- differences in the formation of complex species.

These differences form the basis of the separation procedures by fractional crystallisation, fractional precipitation and ion exchange and solvent extraction.

Also, some trivalent rare earths (Ce, Pr, Tb) may also become tetravalent, and some others (Sm, Eu, Yb) may become divalent. This creates the possibility to use selective oxidation or reduction of these elements to separate them, because the divalent and tetravalent state the elements show marked differences in behavior from the trivalent state (Gupta and Krishnamurthy 2005).

4.3.1 Selective Oxidation

Cerium can be removed from the rare earths mixture after oxidation of naturally occurring Ce(III) to Ce(IV). This valence change occurs, for instance, when bastnaesite is heated in air at 650 °C, or when the rare earth hydroxides are dried in air at temperatures of 120–130°. In aqueous hydroxide suspensions, oxidation can be achieved by chlorination or electrolysis. Also application of ozone is possible for cerium removal. The ozone causes an oxidation precipitation process.

Once oxidized, the Ce(IV) can be removed from the trivalent rare earths in the hydroxide-oxide mixture in two ways:

(1) selective dissolution of trivalent species with dilute acid, or
(2) complete dissolution in concentrated acid

Subsequently, selective precipitation is carried out (Gupta and Krishnamurthy 2005).

However, even the precipitation of 99.8 pure ceric salts does obviously not remove all the cerium from the solution, and additional steps are necessary. A very effective way is air oxidation and solvent extraction. It has become apparent that TBP (tributyl phosphate, see also in Sect. 4.2.3.1) is the best extractant for large scale operations (Gupta and Krishnamurthy 2005).

Although Ce(IV) is stable in aqueous solutions, Pr(IV) and Tb(IV) are not. They cannot be recovered this way. An effective method involves making a solution of rare earth hydroxides in fused potassium hydroxide, and oxidizing this by anodic oxygen or by potassium chlorate (Gupta and Krishnamurthy 2005).

4.3.2 Selective Reduction

It is easily possible to remove Sm(III), Eu(III) and Yb(III) from the trivalent rare earths mixture by reducing them to the divalent state. However, they have to be enriched in the mixture, since they naturally are much less abundant than cerium.

On an industrial scale, a method developed by H.N.McCoy[10] has found acceptance. In a chloride solution, Eu(III) is reduced to Eu(II) by zinc. The divalent Eu is recovered as a sulphate. Zinc does not reduce Sm or Yb (Gupta and Krishnamurthy 2005).

Another method is the reduction of Eu(III) by a photochemical method, followed by the precipitation as a sulphate or chloride. In this method, lanthanide perchlorates and K_2SO_4 (0.05 M) in 10 % isopropanol are used.

Qiu et al. (1991) used also a photochemical method applied to a solution mixture of $SmCl_3$, $EuCl_3$, and $GdCl_3$ in a rare earth saturated ethanol-isopropanol system. Finally $EuCl_2$ was precipitated, which was 92 % pure. Sm(III) and Gd(III) are not reduced (Gupta and Krishnamurthy 2005; Qiu et al. 1991).

Finally, there is a method of separating samarium, europium and ytterbium, based on the fact that the metals Sm, Eu, and Yb cannot be produced by metallothermic reduction[11] of their halides. When a mixture of anhydrous rare-earth halides is reduced with calcium, the halides of Sm, Eu, and Yb are not reduced and remain in the salt slag. They can subsequently be separated from this (Gupta and Krishnamurthy 2005).

---

[10]Herbert Newby McCoy (1870–1945) was an American chemist who taught at the University of Chicago and the University of Utah and was the vice-president of Lindsay Light & Chemical Company. He contributed to numerous papers on physical chemistry, radioactivity and rare earths (Wikipedia 2015b).

[11]This is an extraction technology that produces metal by reaction of one of its compounds with a metallic reducing agent through thermal methods.

4.3.3 Fractional Crystallization

Fractional crystallization can be described as follows: a part of a salt in solution is precipitated by a change in temperature or by evaporation of the saturated solution. If the solubility of the various compounds of the solution differs, the composition of the crystalline precipitate will not be the same as the composition of the original solution. The least soluble substance will be the first to crystallize.

This method has been considered the best of the classical separation procedures for producing individual elements in high purity. The most suitable compounds are ammonium nitrates (for La, Pr, and Nd) and double magnesium nitrates (for Sm, Eu, Gd). Manganese nitrates have also been used for separation of lanthanides of the cerium group (La–Nd). Bromates and sulphates have been used in the separation of the yttrium group (being the heavy lanthanides or HREE)

Erbium, thulium, lutetium and yttrium have been separated by application of a rare-earth hexa-antipyrine iodide salt.[12] Other chemicals applied include a sodium rare earth EDTA[13] salt for separating gadolinium, terbium, dysprosium, and yttrium. For these rare earths, a purity of 99% has been reached with this method.

Fractional crystallization works best for the lanthanum end of the series, as there the differences in ionic radius are the largest. Fractional crystallization is very slow for the heavy rare earths and in the Sm(III)–Gd(III)—region, because the differences in properties between the rare earths decrease as the atomic number increases (Gupta and Krishnamurthy 2005).

Fractional crystallization is one of the oldest methods for the separation of rare earths and is now not used anymore for separation of rare earths (Habashi 2013).

4.3.4 Fractional Precipitation

Fractional precipitation means the removal of part of the dissolved elements from solution by adding a chemical reagent to form a new less-soluble compound. In this respect, it is different from fractional crystallisation, where NO other compound is introduced in the solution.

Hydroxides and double sulphate have extensively been used, as well as double chromate. The latter one especially has been used for the separation of yttrium from the other rare earths. The addition of sodium sulphate to the rare earth solution leads to the precipitation of double sulphates. The elements La, Ce, Pr, Nd and Sm form poorly-dissolvable double sulphates, whereas Ho, Er, Tm, Yb, Lu and Y form well-dissolvable double salts. The salts of Eu, Gd and Dy form salts of intermediate solubility. Generally, the use of this method crudely separates the rare earth mixture

[12]Antipyrine: a white crystalline powder with the formula (CH)(CH)CHNO, which was formerly used to relieve pain and reduce fever.

[13]EDTA is ethylene-diamine-tetra-acetic-acid, $((HO_2CCH_2)_2NCH_2CH_2N(CH_2CO_2H)_2)$.

into three groups, although separation of La and Y is very well possible (Gupta and Krishnamurthy 2005).

4.3.5 Ion Exchange

This method involves the exchange of ions between two electrolytes or between an electrolyte solution and a usually solid material called an ion exchanger. The latter is often a resin, or it is a mineral substance, like a zeolite[14] (Slater 1991; Zagorodni 2007).

Generally, an ion with a higher charge will replace an ion with a lower charge. When the ions have the same charge, the ion with the larger radius displaces the one with the smaller radius according to the laws of mass action. In ion exchange, there is an adsorption stage and an elution stage. In the first, the ions from the solution get loaded on the exchanger, and in the latter, as a consequence of a change of conditions, the ions desorb from the exchanger and go again into solution (Gupta and Krishnamurthy 2005).

For the rare earths, this was first attempted in the late 19th century (1893) for the separation of yttrium and gadolinium using activated carbon. Later the separation of the rare earths was also tried using other exchangers. At first ion exchange seemed not very promising, as the affinity of the chemically closely related rare earth elements for the applied exchangers was not sufficiently different to result in a satisfactory separation. The separation factors for adjacent rare earths were found to be close to unity (Gupta and Krishnamurthy 2005).

The situation changed with the appearance of complexing agents, which could enhance separation factors. Examples of these are citric acid–ammonium citrate, nitrilotriacetic acid (NTA) or ethylene-diamine-tetra-acetic acid (EDTA). Exchange with the citrate worked, but the application of NTA and EDTA did not lead at first to satisfactory results. Later, after the invention of the so-called band-displacement technique (Spedding et al. 1954), separation improved. The most useful commercial complexing agents are EDTA and HEDTA[15] (Gupta and Krishnamurthy 2005).

EDTA can effectively separate most of the rare earths from each other. Only for the pairs Eu–Gd, Dy–Y, and Yb–Lu does it not work so well. Separation of Dy–Y with HEDTA is, however, possible. HEDTA also shows good results for Tm–Yb–Lu, and for La–Ce–Pr–Nd–Y–Sm, and Ho–Er–Tm–Yb–Lu mixtures.

---

[14] *Definition*: A zeolite mineral is a crystalline substance with a structure characterized by a framework of linked tetrahedra, each consisting of four O-atoms surrounding a cation. This framework contains open cavities in the form of channels and cages. These are usually occupied by H_2O molecules and extra framework cations that are commonly exchangeable. The channels are large enough to allow the passage of guest species (Coombs et al. 1997).

[15] HEDTA is hydroxyethyl-ethylene-diamine-triacetic acid.

In ion-exchange separations, temperature plays a crucial role. Some separations (e.g., Gd–Eu and Eu–Sm) worked very well at, for instance, 92 °C, whereas at room temperature they encountered difficulties (Gupta and Krishnamurthy 2005).

The band-displacement technique with ammonium-ethylene-diamine-tetra-acetate as the complexing agent has made ion exchange the most-used commercial process for rare-earth-element separation since the mid 1950. The advantages are many: separation of a mixture of 15 rare earths into its components with 99.99 % purity in one pass through the system; potential upscaling to multi-ton quantities; the possibility for recovery and recycling of water; and retention of ion and complexant. The technique is still used today, even with the possibility of solvent extraction, which is discussed hereafter (Gupta and Krishnamurthy 2005).

4.3.6 Solvent Extraction

Solvent extraction (SX), also called **liquid-liquid-extraction** (LLE), is the transfer of one or more solutes contained in a feed solution to another, essentially ***immiscible*** liquid (solvent). It takes place with aqueous and organic solutions at ambient temperatures and pressures (Kislik 2012).

The process was first applied to rare earths in the 1930s. Separation of the rare earths by solvent extraction is currently the most favored method to obtain very pure fractions of every rare earth. One of the liquid phases is an aqueous solution and the other is a non-aqueous phase (usually an organic liquid). There are many advantages to the use of solvent extraction for the separation of the rare earths. A major advantage is that the rare earth loading in the solvent or extractant can be very high. Aqueous solutions with up to 140 g of rare earth oxide per liter can be used (Gupta and Krishnamurthy 2005). Another advantage is that, with relatively simple equipment and rather quickly, very high purity can be achieved (Kislik 2012). Purities of >99.999 %, reached almost exclusively with solvent extraction, are reported in Gupta and Krishnamurthy (2005).

The organic phase used in the extraction process usually consists of two or more substances. One is the extractant itself, but very often this is as such a very viscous material, which cannot be applied in practice. It is therefore dissolved into a suitable solvent to ensure that there is good contact with the aqueous phase.

Usually the complete transfer of the metal ions to the organic phase does not proceed to its full extent in one contact. Therefore the process has to be repeated several times. After this, scrubbing and stripping must take place. This means that the loaded organic phase is brought in contact with an aqueous solution to collect the impurities extracted by the solvent. Stripping means that the scrubbed organic phase is brought in contact with an aqueous phase, in order to recover the main extracted substance(s) from the organic phase back to the aqueous phase (Gupta and Krishnamurthy 2005).

4.4 Scandium

Scandium is also extracted by solvent extraction from its resources. These are, however, essentially non-rare earth minerals. The main scandium resource minerals are uranium minerals, and trace amounts occur in iron and magnesium rich rocks.

Therefore, scandium is produced during the extraction of uranium. Initially, it is extracted together with uranium, but it is collected during the subsequent purification of uranium (Gupta and Krishnamurthy 2005).

References

Coombs DS, Alberti A, Armbruster T, Artioli G, Colella C, Galli E, Grice JD, Liebau F, Mandarino JA, Minato H, Nickel EH, Passaglia E, Peacor DR, Quartieri S, Rinaldi R, Ross MI, Sheppard RA, Tillmanns E, Vezzalini G (1997) Recommended nomenclature for zeolite minerals; report of the subcommittee on zeolites of the international mineralogical association, commission on new minerals and mineral names. Can Mineral 35:1571–1606

Encyclopedia Brittanica (2015a) Froth flotation (ore dressing). http://www.britannica.com/EBchecked/topic/210944/flotation. Accessed Mar 2015

Encyclopedia Britannica (2015b) Tributyl phosphate. http://www.britannica.com/science/tributyl-phosphate)

Gupta CK, Krishnamurthy N (2005) Extractive metallurgy of the rare earths. CRC Press, 484 p

Habashi F (2013) Extractive metallurgy of rare earths. Can Metall Q 52(3):224–233

Kislik VA (2012) Solvent extraction: classical and novel approaches. Elsevier, 555 p

Lapin AV, Lyagushkin AP (2014) The kovdor apatite-francolite deposit as a prospective source of phosphate ore. Geol Ore Deposits 56(1):61–80

Mindat.org (2015) Leucoxene. (http://www.mindat.org/min-6174.html). Accessed Jan 2015

Qiu LF, Kang X-H, Wang T-S (1991) A study on photochemical separation of rare earths: the separation of europium from an industrial concentrate material of samarium, europium, and gadolinium. Sep Sci Technol 28(2):199–221

Slater MJ (1991) Principles of ion exchange. Butterworth–Heinemann Ltd., Oxford, 182 p

Spedding FH, Powell JE, Wheelwright EJ (1954) The use of copper as the retaining ion in the elution of rare earths with ammonium ethylene diamine tetra acetate solutions. J Am Chem Soc 76:2557–2560

Wall F, Mariano AN (1996) Rare earth minerals in carbonatites: a discussion centred on the Kagankunde Carbonatite, Malawi. In: Jones AP, Wall F, Williams CT (eds) Rare earth minerals—chemistry, origin and ore deposits. The Mineralogical Society Series, vol 7. Chapman and Hall, pp 193–225

Wikipedia (2015a) Pyrophosphate. http://en.wikipedia.org/wiki/Pyrophosphate. Accessed Mar 2015

Wikipedia (2015b) Herbert Newby McCoy http://en.wikipedia.org/wiki/Herbert_Newby_McCoy. Accessed Mar 2015

Wills BA, Napier-Munn TJ (2006) Mineral processing technology. In: An introduction to the practical aspects of ore treatment and mineral recovery, 7th edn. Elsevier—Butterworth & Heinemann, 444 p

Zagorodni A (2007) Ion exchange materials—properties and applications. Elsevier Ltd., 477 p

Zhang P, Kejie T, Yang Z, Yang X, Song R (2002) Rare earths, niobium and tantalum minerals in Bayan obo ore deposit and discussion on their genesis. J Rare Earths 20(2):81–86

Chapter 5
Applications of the Rare Earths

Abstract This chapter gives an overview of the most important applications of the rare earth elements. The number applications discussed is large, but not exhaustive. Among those treated are catalysts, metal alloys, HT-superconductors, batteries, phosphors, glass, glass-polishing agents, permanent magnets, pigments, nuclear-control rods, photographic filters, lasers, and the scintillation crystals used in PET-scanners. The permanent magnets have themselves a large number of different applications, which are listed in this chapter.

5.1 Introduction

Many technological items that people are familiar with in modern society depend critically on the rare earths. But these elements are used also in other applications that people do not see directly. The number of applications discussed is large, but not exhaustive in keeping with the preferred brevity and conciseness of volumes in the series in which this book appears.

The applications may be very high tech, but not necessarily. Examples of the latter are for instance pigments or the alloy called **misch metal** (see Sect. 5.5).

In Table 5.1. you will find the most important applications for a number of elements.

5.2 Scandium

Scandium is used mainly as an alloying metal for aluminum (Scandium.org 2015). A minor application is in metal halide light bulbs (Gendre 2003). Another minor application for the radioactive isotope scandium-46, with a half life of 84 days, is in oil refineries to monitor the movement of fractions during the refining process.

J.H.L. Voncken, *The Rare Earth Elements*, SpringerBriefs in Earth Sciences,
DOI 10.1007/978-3-319-26809-5_5

Table 5.1 Usage of rare earth elements by application, given in % (Curtis 2010)

Application	La	Ce	Pr	Nd	Sm	Eu	Gd	Tb	Dy	Y	Other
Magnets			23.4	69.4			2	0.2	5		
Battery alloys	50	33.4	3.3	10	3.3						
Metallurgy exclusive of batteries	26	52	5.5	16.5							
Automotive catalysts	5	90	2	3							
FCC-catalysts[a]	90	10									
Polishing powder	31.5	65	3.5								
Glass additives	24	66	1	3						2	4
Phosphors	8.5	11				4.9	1.8	4.6		69.2	
Ceramics	17	12	6	12						53	
Other	19	39	4	15	2		1			19	

[a]FCC catalysts are fluid catalytic-cracking catalysts

Dilute solutions of scandium sulfate, when applied to corn, peas and wheat, increase the number of seeds which germinate successfully (Emsley 2001).

Scandium is used on a limited scale to produce high-intensity lights. The radioactive isotope Sc^{40} is used as a tracing agent in, for instance, refinery crackers for crude oil. When added to mercury-vapor lamps, scandium iodide produces a light source which resembles daylight (Hammond 2015).

5.3 Yttrium

The applications of yttrium are diverse. It is used in alloys (e.g., with magnesium, chromium, molybdenum and zirconium). Yttrium YVO_4-Eu and Y_2O_3-europium phosphors are the standard red components in color televisions and monitors (Hammond 2015). Yttrium oxide is used in lenses for cameras to make the glass heat and shock resistant.

The microwave and radar applications are in the form of yttrium-iron garnet (YIG) and its aluminum counterpart yttrium-aluminum garnet (YAG), which is used for lasers.

One of the first high-temperature superconductors (superconducting at −183 °C) discovered was $YBa_2Cu_3O_{7-9}$ (Wu et al. 1987) (Fig. 5.1).

Yttrium-aluminum garnet, with a hardness of 8.5, is used as a gemstone (imitation diamond). In small amounts (0.1–0.2 %), yttrium can be added to chromium, molybdenum, zirconium and titanium to the reduce grain size of these metals. Magnesium and aluminum alloys possess increased strength when small amounts of yttrium are added. Also the metal is used as a deoxidizer for vanadium and other non-ferrous metals (Hammond 2015).

Fig. 5.1 Magnet levitating above an yttrium-containing superconductor cooled with liquid nitrogen. *Source* Mai-Linh Doan, via Wikimedia Commons, under license CC BY-SA 3.0. No changes made. Link: http://commons.wikimedia.org/wiki/File:Meissner_effect_p1390048.jpg

Yttrium-iron garnet (YIG), made in large crystals, can be used in optical applications, and if deposited as a thin layer, it can be used for magnetic recording. It can also be used as a transducer of acoustic energy (Emsley 2001).

Yttrium is also widely used in phosphors for lamps and displays in compounds such as YVO_4: Eu and Y_2O_3 (Hammond 2015).

5.4 Lanthanum

Lanthanum, often occurring together with cerium, belongs to the most abundant rare earth elements. The applications of lanthanum are described below.

5.4.1 Alloy

The metal lanthanum has no commercial uses of its own. In alloys, there are several applications. The most well-known alloy is misch metal (see at Cerium, Sect. 5.5, for a detailed description).

5.4.2 Batteries

Another alloy is lanthanum-nickel-hydride. NiMH batteries are an alkaline storage battery due to the use of potassium hydroxide (KOH) as the electrolyte. The batteries are rechargeable and form a dominant factor in the market. A highly-used variety is the LaNiH-battery. This battery is superior to the well-known Ni–Cd-containing batteries. Technically, they are an extension of the Ni–Cd-batteries, with a hydrogen absorbing negative electrode replacing the cadmium based electrode. Considering the toxicity of cadmium, this is an important improvement (Royal Society of Chemistry 2015).

5.4.3 Catalyst

Lanthanum is also very important in fluid catalytic cracking or FCC (Yung and Bruno 2012). FCC is the most important conversion process in the petrochemical industry. A long chain of CH-bonds is converted to short hydrocarbon compounds by means of a catalyst (Fig. 5.2).

5.4.4 Special Optical Glasses

Lanthanum(III) oxide (La_2O_3) is used for making special optical glasses. It improves the optical properties and alkali resistance of the glass (Vinogradova et al. 2004; Royal Society of Chemistry 2015).

Fig. 5.2 Petrochemical plant in Saudi Arabia. *Source* "TASNEE 001" by Secl. Licensed under CC BY 3.0 via Wikimedia Commons. Link: http://commons.wikimedia.org/wiki/File:TASNEE_001.jpg#/media/File:TASNEE_001.jpg

5.4.5 *Superconductor*

La is also a key element in the very first high-temperature ceramic superconductor $Ba_xLa_{5-x}Cu_5O_{5(3-y)}$ with x = 1 and 0.75 and y > 0 (Bednorz and Müller 1986).

5.4.6 *Phosphors*

Lanthanum has been used extensively in several types of phosphors (e.g., Lehmann and Isaacs 1978).

5.5 Cerium

Cerium is one of the most abundant rare earth elements. Ceria (CeO_2) is the most widely used compound of cerium (Royal Society of Chemistry 2015). Below the most important applications are described.

5.5.1 *Polishing Compound*

Ceria (CeO_2) is the most widely used compound of cerium. The main application of ceria is as a polishing compound, e.g., chemical-mechanical planarization (CMP[1]). It is used to produce high-quality optical surfaces (Fig. 5.3).

5.5.2 *Fluid Catalytic Cracking*

Cerium(IV)-oxide, CeO_2, (see Fig. 5.4) is, like La(III)-oxide, used in fluid catalytic cracking.

5.5.3 *Catalytic Converter*

In automotive catalytic converters, Ce(III)oxide is used as a catalyst for the oxidation of CO and NO_x. See Fig. 5.5 for a schematic image of a catalytic converter.

---

[1]Chemical mechanical polishing/planarization is a process of smoothing surfaces with the combination of chemical and mechanical forces. It can be thought of as a hybrid of chemical etching and free abrasive polishing.

Fig. 5.3 Cerium oxide polishing powder. Image used with Permission. Photo credit: Mama's Minerals, Inc.—http://mamasminerals.com

Fig. 5.4 Cerium(IV) oxide. *Source* Wikipedia (2015c) Cerium (IV)-oxide. http://en.wikipedia.org/wiki/Cerium(IV)_oxide. Public domain image

5.5.4 Component in Special Glass

Cerium is also used in glass that is subject to alpha, gamma, X-ray, light and electron radiation. The cerium decreases the rate of discoloration of the glass, primarily by preventing the formation of divalent Fe (Emsley 2001).

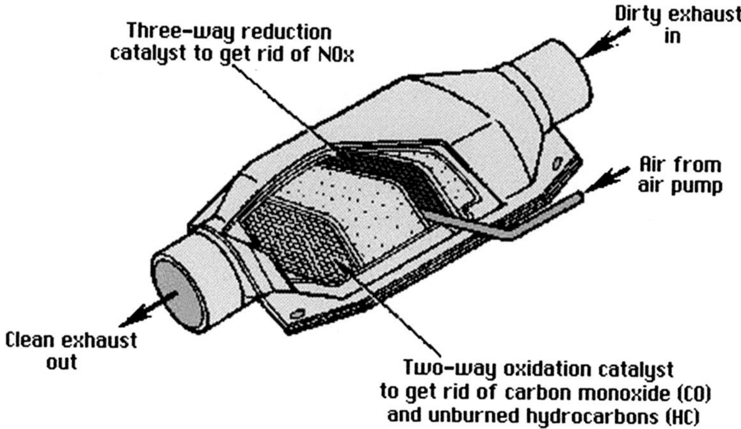

Fig. 5.5 Schematic image of a catalytic converter. *Image source* http://benature.tv/catalytic-converters-are-the-achievement-for-human-because-of-the-advancement-of-the-science-and-technology/. Image used with permission

5.5.5 Alloys

One of the most important alloys in which cerium is an important component is the so-called *misch metal* (Wikipedia 2015a) (from the German, "Mischmetall" meaning "mixed metal"). It is also called cerium misch metal, or rare earth misch metal. The composition is not fixed and variations occur. A typical composition includes approximately 50 % cerium, 25 % lanthanum, 15 % neodymium, and 10 % other rare-earth metals (praseodymium, europium, samarium) (Encyclopaedia Britannica online 2015). In Fig. 5.6, an image is shown of misch metal.

Fig. 5.6 Misch metal. *Source* "Misch metal" by Spypredator—Licensed under CC BY-SA 3.0 via Wikimedia Commons (Wikipedia 2015b). Link: http://commons.wikimedia.org/wiki/File:Mischmetal.JPG#/media/File:Mischmetal.JPG. No changes made

This is a so-called pyrophoric alloy. The most common use of misch metal is in the ignition devices of lighters and torches, where it is commonly called "*flint*". The material is blended with iron oxide and magnesium oxide in order to produce a harder material. This material is also known as ferrocerium. Misch metal was invented by Carl Auer Baron von Welsbach (1858–1929) and patented in 1906 (Patent US837017).

Cerium dioxide is also used to opacify enamels (Yuriditsky 2003), for photochromic glasses (e.g. Smith 1967), for ceramic coatings (e.g. Nazeri et al. 1997), in refractory oxides (ceria stabilized zirconia, e.g. Jones 1997), cathodes for solid oxide fuel cells (Kharton et al. 2001), capacitors (Cheng et al. 2010) and semiconductors (e.g. Preisler 2003).

5.5.6 Pigment

Until a few years ago, the red pigment for coloring containers, toys, household wares, plastic crates, etc. was Cadmium Red. This is *Cadmium sulfoselenide red*, $C_9H_{13}CdN_2O_4SSe$ (CAS[2]: 58339-34-7), also known as Pigment Red 108 (see Chemnet 2015). The cadmium compound is now considered environmentally undesirable, so it is being replaced by the non-toxic *Cerium(III) sulphide* (Ce_2S_3; CAS 12014-93-6). This compound gives a rich red color and is stable up to 350 °C (Emsley 2001).

5.5.7 Other

Other uses of cerium are in chromium plating, where a little cerium is added to the electrolyte solution of the plating baths. It helps in the process to prevent the formation of Cr(III)-ions, which electrolytically are difficult to reduce to the metal (Emsley 2001).

Cerium oxide is an important constituent of incandescent gas mantles. Its use as hydrocarbon catalysts in "self-cleaning ovens" is becoming more and more important (Hammond 2015).

Cerium is also used as an activator in yttrium-silicate phosphors (e.g. Muresan et al. 2010).

---

[2]CAS: Chemical Abstracts Service is a division of the American Chemical Society. Entering the CAS registry number in Google, e.g., leads directly to the specific compound.

5.6 Praeseodymium

Like the other lanthanides, praseodymium has several applications. One of the oldest is in misch metal, wherein it constitutes approximately 5 % of the composition.

Other major uses of praseodymium are:

- Alloying element with magnesium for producing high-strength metals for the aircraft industry (e.g., Rokhlin 2003).
- Pigment for glass and enamel (green, yellow). See, for instance, Del Nero et al. (2004).
- Praseodymium-doped glass, called didymium glass (Avalon 2015a), turns yellow and is used in welding goggles because it blocks infrared radiation.
- Praseodymium is also applied as a catalyst (e.g. Asami et al. 1997).
- Praseodymium is used for fluoride-glass optical fibres (Jha et al. 1995).

5.7 Neodymium

Neodymium is most well-known for the very strong permanent magnets that are made of it: neodymium-iron-boron magnets, also named neodymium magnets, neo-magnets, or NdFeB-magnets Herbst and Croat (1991). These are strong permanent magnets with the chemical formula $Nd_2Fe_{14}B$. They were developed in 1982 simultaneously by General Motors and Sumitomo Special Metals. These magnets are the strongest type of permanent magnet commercially available and can be bought in sintered and bonded forms. Their most important drawbacks are the rather low maximum operating temperature compared with SmCo magnets, and the fact that they corrode easily. They are therefore usually coated. Both single-component metal coatings such as nickel, chromium, aluminum, zinc, tin, silver, gold, as well as the multi-component, multilayered ones like Ni–Cr or Ni–Cu, are used (Drak and Dobrzański 2007) (Table 5.2).

Table 5.2 Properties of sintered and bonded NdFeB-magnets (MMC Magnets 2015)

Magnet type	Remanence (mT) Depending on the variety	Intrinsic coercive force, Hci (kA/m) Depending on the variety	Maximum operating temperature (°C) Depending on the variety
$Nd_2Fe_{14}B$ (sintered)	1050–1450	877–2786	80–220
$Nd_2Fe_{14}B$ (bonded)	140–620	143–1200	130–180

Following are listed applications of neodymium magnets (Brown et al. 2002).

- Computer and office automation
- Disk-drive spindle motors and voice coil motors
- CD-ROM spindle motors and pick-up motors
- Printer and fax stepper motors
- Printer hammers
- Copy machine rollers
- Automotive industry
- Starter motors
- Electric steering
- Sensors
- Electric fuel pumps
- Instrumentation gauges
- Brushless DC motors
- Actuators
- Alternators
- Consumer electronics
- VCRs and camcorders
- Cameras
- Speakers, headsets
- Microphones
- Pagers
- DVD players
- Watches
- Cell phones
- Appliances
- Portable power tools
- Household appliance motors
- Scales
- Air conditioners
- Water pumps
- Security systems
- Factory automation
- Magnetic couplings
- Pumps
- Motors
- Servo motors
- Generators
- Bearings
- Medical industry
- MRI
- Surgical tools
- Therapeutic implants.

A well-known method to increase the coercivity in Nd–Fe–B magnets is to substitute the heavy rare-earth (HREE) elements like Dysprosium (Dy) and Terbium (Tb) for Nd (Dhakal et al. 2014; Hirosawa et al. 1986; Hirota et al. 2006). Substitution with dysprosium also raises the Curie-point for these magnets, so that they can be used at higher temperatures. The amount of dysprosium in neodymium magnets may vary from 0.8 to 1.2 % (NdFeB-info.com 2015).

Neodymium is also used as a dopant in yttrium-aluminum garnet (YAG) lasers or Nd:YAG lasers (Geusic et al. 1964). The chemical formula of YAG is $Y_3Al_5O_{12}$.

5.8 Promethium

The radioactive element promethium has a very short half-life and very few applications outside of the laboratory. Most of it is used exclusively in research. Promethium can also be used as a source of X-rays and radioactivity in measuring instruments.

5.9 Samarium

The most important application of samarium is in magnets. Samarium-cobalt magnets were discovered at the end of the 1960s (Strnat et al. 1967). There are two types of these magnets: $SmCo_5$ and Sm_2Co_{17}. The magnets are characterized by a high strength, high maximum-operating temperature and high coercivity.[3] Several varieties are made from both compositions. The magnets are brittle and so crack easily. On the other hand, they have very good resistance to corrosion and thus do not need any protective coating (in contrast to neodymium magnets) (Table 5.3).

Samarium has, however, also other applications. Samarium(II)Iodide (SmI_2), also known as Kagan's reagent, is a powerful reductor, and can easily reduce water to hydrogen. It is used in organic synthesis as a reducing or coupling agent, together with ytterbium(II)iodide, YbI_2 (Girard et al. 1980).

Samarium perchlorate (Sm(III)oxide in perchloric acid) is a particularly suitable calibration material for checking the wavelength scale of a spectrophotometer over the most commonly used range of 200–500 nm, as it has peaks throughout this region (Optiglass 2015).

[3]Magnetic coercivity: measure of the potential of a ferromagnetic material to resist an external magnetic field without becoming demagnetized.

Table 5.3 Properties of SmCo-magnets (MMC Magnetics 2015)

Magnet type	Remanence, B_r (mT) depending on the variety	Intrinsic coercive force, Hci (kAm) depending on the variety	Maximum operating temperature (°C) depending on the variety
$SmCo_5$	820–950	>1270 to >1590	250
Sm_2Co_{17}	1000–1150	>1272 to >2782	375

5.10 Europium

Europium has not many applications. Commercial applications of europium almost always take advantage of the phosphorescence, either in the 2+, or 3+ oxidation state. Industrial application started in the 1960s (Wickersheim and Lefever 1964). Many advances have been made since, leading to the discovery of europium activated yttrium oxysulfide ($Y_2O_2S:Eu^{3+}$). This is a red phosphor. Other main industrial phosphors doped with europium are: $Sr_5(PO_4)_3Cl:Eu^{2+}$ and $BaMgAl_{11}O_{17}:Eu^{2+}$ for blue, and $Y_2O_3:Eu^{3+}$ for red. Green phosphors involve terbium (Tb^{3+}) (Caro 1998; Nazarov et al. 2004). The phosphors are for instance used in flat screen monitors and televisions.

5.11 Gadolinium

Gadolinium has many very specialized uses. None of them are, however, large scale. Gadolinium is used in shielding of nuclear reactors, and in neutron imaging (neutron radiography). Metallurgical applications of gadolinium are the improvement of the workability and of the resistance of iron, chromium, and related alloys to high temperatures and oxidation.

Gadolinium is ferromagnetic at room temperature, with a Curie point[4] of 293.4 K or 20.25 °C (Geschneider 2005).

This property makes gadolinium useful in magnetic resonance imaging[5] (MRI). In this procedure, solutions of organic gadolinium complexes and compounds are used intravenously to enhance contrast in the images. A common Gd(III) contrast agent is "Magnevist" or Gadpentetic acid[6] (Sherry et al. 2009).

[4]The Curie point is the temperature at which the permanent magnetism of a material changes to paramagnetism, which is only present in a magnetic field.

[5]Magnetic resonance imaging (MRI) is a medical imaging technique using strong magnetic fields, but no ionizing radiation. It is widely used for imaging soft human tissues.

[6]Gadopentetic acid is a trade name of a complex of gadolinium with DTPA—diethylenetriaminepentacetate (Sherry et al. 2009).

5.12 Terbium

Terbium is used to dope calcium fluoride, calcium tungstate, and strontium molybdate. Their application is in luminescent materials (e.g., Li et al. 2009; Nazarov et al. 2004). It also can be used together with zirconia (ZrO_2) as a crystal stabilizer for solid-oxide fuel cells that operate at high temperature (Hammond 2015).

5.13 Dysprosium

Dysprosium is a common addition to NdFeB-magnets. It increases intrinsic coercivity of magnets whilst reducing remanence. It also raises the Curie-point of the magnets. Combined addition of Dy and Nb increases the coercivity, as $Dy_2Fe_{14}B$ is known to have a stronger magnetocrystalline anisotropy than $Nd_2Fe_{14}B$ (Kim et al. 1994). However, excessive Dy weakens the remanence due to the lower *J*s (spin angular momentum) of $Dy_2Fe_{14}B$. The optimum amount of Dy is 0.5 atom percent (Chen et al. 2010).

A dysprosium oxide-nickel cermet[7] is used in cooling nuclear-reactor rods. This cermet absorbs neutrons readily without swelling or contracting under prolonged neutron bombardment Another field of application is dosimeters for radioactive exposure (Emsley 2001).

Other uses of dysprosium are as a dopant in $BaTiO_3$, which is used to make capacitors of small dimensions but with high capacity, and temperature compensating capacitors (Emsley 2001).

5.14 Holmium

Holmium, having the highest magnetic field strength of any element, is used as a magnetic flux concentrator in pole pieces of high-strength magnets (Hoard et al. 1985). Holmium is also used in nuclear reactors to keep the chain reaction in check, because it absorbs neutrons (Emsley 2001). The magnetic properties also make it useful for application in YIG (yttrium-iron-garnet)[8]. YIGs are, for instance, used in microwave equipment.

The wavelength of holmium lasers is 2.08 microns (safe to the human eye), which makes it safe to use them in a variety of medical and dental applications. For the latter, holmium-doped yttrium-aluminum garnet (YAG) and yttrium-lanthanum-fluoride (YLF)[9] lasers are used (Gupta and Krishnamurthy 2005).

---

[7]A cermet is a composite material composed of ceramic (cer) and metallic (met) materials.

[8]Yttrium-iron garnet: $Y_3Fe_5O_{12}$.

[9]YLF: $LaYF_4$.

Fig. 5.7 Holmium(III) oxide. *Left* natural light, *right* fluorescent-lamp light. File from Wikimedia Commons, the free media repository. Licensed under CC BY 3.0 via Wikimedia Commons. http://commons.wikimedia.org/wiki/File:Holmium(III)_oxide.jpg. Original file uploaded by Filousoph

Holmium is used in glass for spectrophotometer calibration (Allen 2007). Also, holmium is used as a colorant for cubic zirconia (imitation diamond) and glass, which yields yellow or red colors (Fig. 5.7).

5.15 Erbium

Erbium is commonly used as a photographic filter (Avalon 2015b).

Erbium oxide is important in eyeglasses for absorption of visible and infrared light. Safety goggles for welders and glassblowers contain erbium oxide. Erbium oxide has a pink color and has been used as a pigment to color porcelain and enamels glazes (Emsley 2001).

In optical fibers, erbium is doped at regular intervals in order to amplify signals. This works as follows: erbium converts other wavelengths that are sent through the fiber to the wavelength of the signal that carries the information (Emsley 2001).

Erbium is also applied in phosphors that can convert infrared light into visible light. This results in a green image (Emsley 2001).

Lasers with erbium work at a wavelength of 2.9 μm, which is strongly absorbed by water. In this way energy can be delivered without overheating. Such lasers are used in medical and dental applications (Emsley 2001).

The isotope Er-167 is highly neutron absorbing and therefore is used in the production of special nuclear-fuel rods (Emsley 2001).

The metallurgical applications of erbium are few, as it slowly tarnishes in air, and it is attacked by water. However, it is added to alloys with metals such as vanadium, because it lowers their hardness. The alloys become more workable because of this (Emsley 2001).

5.16 Thulium

Thulium is one of the most expensive of the rare earths, and therefore has little applications (Emsley 2001). Thulium is used as a dopant in YAG and YLF. Efficient flash lamp and laser diode-pumped laser operation has been achieved in Tm^{3+}:YAG and Tm^{3+}:YLF co-doped either with Cr^{3+} or Ho^{3+}. Thulium is also used in thulium-doped holmium lasers, which have outputs in the 2 μm region, a wavelength range of interest for coherent radar systems, remote sensing and medical applications (Koechner 2006).

5.17 Ytterbium

Ytterbium has little commercial applications, as it is a quite rare element. World production is around 50 tons per year. Beneficial uses are the strengthening of steel, doping of phosphorceramic capacitors, other electronic devices, and it can even be used as an industrial catalyst. As ytterbium has a single absorption band at 985 nm in the infrared, this has be used to convert radiant energy into electrical energy in equipment that couple it to photocells (Emsley 2001).

The ultra-stable atomic clock of the US National Institute of Standards and Technology (NIST) with an instability of 10^{-18}, utilizes spin-polarized, ultra-cold atomic ytterbium (Hinkley et al. 2013).

5.18 Lutetium

Lutetium is also a very rare metal, with a world production (as lutetium oxide) of approximately 10 tons per year. One commercial application is known: it is used as a β-emitter, (when Lu-176 has been exposed to neutron activation) in the oil refining industry (Emsley 2001). It was used in so-called-bubble memory, a technology for computer memory that has become obsolete, since hard disks have made their rapid substantial advance. It is still used in scintillation crystals for PET-scanners (positron emission scanners). It is used here in the form of cerium-doped Lutetium oxy-orthosilicate (LSO), with the formula $Ce_{2x}\ Lu_2(1 - x)\ SiO_5$, where x is within the range from approximately 2×10^{-4} to approximately 3×10^{-2} (Melcher 1990; Daghighian et al. 1993).

References

Allen DW (2007) Holmium oxide glass wavelength standards. J Res Natl Inst Stand Technol 112:303–306

Asami K, Kusakabe K, Ashi N, Ohtsuka Y (1997) Synthesis of ethane and ethylene from methane and carbon dioxide over praseodymium oxide catalysts. App Catal A: General 156:43–56

Avalon (2015a) http://avalonraremetals.com/rare_metals/praseodymium/

Avalon (2015b) http://avalonraremetals.com/rare_metals/erbium

Bednorz JG, Müller KA (1986) Possible high T c superconductivity in the Ba–La–Cu–O system. Z Phys B: Condens. Matter 64:189–193

Brown D, Ma B-M, Chen Z (2002) Developments in the processing and properties of NdFeB-type permanent magnets. J. Mag Magnet Mater 248:432–440

Caro P (1998) Rare earths in luminescence. In: Saez R, Caro P (eds) Rare Earths. Editorial Complutense, SA, pp 323–325

Chemnet (2015) http://www.chemnet.com/cas/es/58339-34-7/cadmium%20sulfoselenide%20red.html Accessed March 2015

Chen Z, Luo J, Guo Z (2010) Effect of Dy substitution on the microstructure and magnetic properties of nanograin Nd-Fe-B single-phase alloys. Int J Min Metall Mater 17(3):335–339

Cheng CH, Hsu HH, Chen WB, Chin A, Yeh FS (2010) Characteristics of cerium oxide for metal–insulator–metal capacitors. Solid-State Lett 13(1):H16–H19

Curtis N (2010) Rare earths, we touch them every day. In: Lynas Presentation at the JP Morgan Australia Corporate Access Days, New York, 27–28 Sept 2010

Daghighian F, Shenderov P, Pentlow KS, Graham MC, Eshaghian B, Melcher CL, Schweitzer JS (1993) Evaluation of cerium doped lutetium oxy-orthosilicate (LSO)scintillation crystal for PET. IEEE Trans Nucl Sci 40(4):1045–1047

Del Nero G, Cappelletti G, Ardizzone S, Fermo P, Gilardoni S (2004) Yellow Pr-zircon pigments—the role of praseodymium and of the mineralizer. J Eur Ceram Soc 24:3603–3611

Dhakal DR, Namkung S, Lee M-W, Jang T-S (2014) Effect of dysprosium-compounds treatment on coercivity of Nd-Fe-B sintered magnets. Curr Nanosci 10:28–31

Drak M, Dobrzański LA (2007) Corrosion of Nd-Fe-B permanent magnets. J Achiev Mater Manuf Eng 20(1–2):239–242

Encyclopaedia Britannica online (2015) http://www.britannica.com/technology/misch-metal

Emsley J (2001) Nature's building blocks: an A-Z guide to the elements. Oxford University Press, Oxford, 538 pp

Gendre MF (2003) Two centuries of electric light source innovations. URL: http://www.einlightred.tue.nl/lightsources/history/light_history.pdf, 12 pp

Geschneider Jr KA (2005) Physical properties of the rare Earth metals. In: Lide DR (ed) CRC handbook of chemistry and physics, 95th edn. CRC Press, Boca Raton, FL, 4.118

Geusic JE, Marcos HM, van Uitert LG (1964) Laser oscillations in Nd-doped yttrium aluminum, gallium and gadolinium garnets. Appl Phys Lett 4(10):182–184

Girard P, Namy JL, Kagan HB (1980) Divalent lanthanide derivatives in organic synthesis. 1. Mild preparation of SmI_2 and YbI_2 and their use as reducing or coupling agents. J Am Chem Soc 102(8):2693–2698

Gupta CK, Krishnamurthy N (2005) Extractive metallurgy of the rare earths. CRC Press, Boca Raton, p 31

Hammond CR (2015) The elements. In: Haynes WM, Bruno TJ, Lide DR(eds) CRC handbook of chemistry and physics, 96th edn, Internet Version. CRC Press, Boca Raton, Section 4, pp 1–36

Herbst JF, Croat JJ (1991) Neodymium-iron-boron permanent magnets. J Magn Magn Mater 100:57–78

Hinkley N, Sherman JA, Phillips NB, Schioppo M, Lemke ND, Beloy K, Pizzocaro M, Oates CW, Ludlow AD (2013) Science express, 22 Aug 2013, 3 pp

Hirosawa S, Matsuura Y, Yamamoto H, Fujimura S, Sagawa M, Yamauchi H (1986) Magnetization and magnetic anisotropy of $R_2Fe_{14}B$ measured on single crystals. J Appl Phys 59(3):873–879

Hirota K, Nakamura H, Minowa T, Honshima M (2006) Coercivity enhancement by the grain boundary diffusion process to Nd-Fe-B sintered magnets. IEEE Trans Magn 42(10):2909–2911

Hoard RW, Mance SC, Leber RL, Dalder EN, Chaplin MR, Blair K, Nelson DH, Van Dyke DA (1985) Field enhancement of a 12.5 T magnet using Holmium poles. IEEE Trans Magn 21 (2):448–450

Jha A, Naftalyt M, Jorderyt S, Samson BN, Taylor ER, Hewakl D, Paynex D.N., Poulain, M., Zhangs, G. (1995) Design and fabrication of Pr^{3+}-doped fluoride glass optical fibres for efficient 1.3 micrometer amplifiers. Pure Appl Opt 4:417–424

Jones RL (1997) Some aspects of the hot corrosion of thermal barrier coating. J Therm Spray Technol 6(1):77–84

Kharton VV, Figueiredo FM, Navarro L, Naumovich EN, Kovalevsky AV, Yaremchenko AA, Viskup AP, Carneiro A, Marques FMB, Frade JR (2001) Ceria-based materials for solid oxide fuel cells. J Mat Sci 36:1105–1117

Kim AS, Camp FE, Stadelmaier HH (1994) Relation of remanence and coercivity of Nd, (Dy)-Fe, (Co)-B sintered permanent magnets to crystallite orientation. J Appl Phys 76:6265–6267

Koechner W (2006) Solid-state laser engineering. Springer Science+Business Media, Inc., p 49

Lehmann W, Isaacs ThJ (1978) Lanthanum and yttrium halo-silicate phosphors. J Electrochem Soc: Solid State Sci Technol 445–448

Li X, Yang Z, Guan L, Guo Q (2009) A new yellowish green luminescent material SrMoO4:Tb^{3+}. Mat Lett 63:1096–1098

Melcher CL (1990) Lutetium orthosilicate single crystal scintillator detector. US Patent 4,958,080

MMC Magnetics. http://www.mmcmagnetics.com/ourproducts/main_SmCo.htm. Accessed Dec 2014

Muresan L, Stefan M, Bica E, Morar M, Indrea E, Popovici EJ (2010) Spectral investigations of cerium activated yttrium silicate blue emitting phosphor. J Optoelectr Adv Mat—Symp 2 (1):131–135

Nazarov MV, Jeon DY, Kang JH, Popovici E-J, Muresan L-E, Zamoryanskaya MV, Tsukerblat BS (2004) Luminescence properties of europium–terbium double activated calcium tungstate phosphor. Solid State Commun 131:307–311

Nazeri PP, Trzaskoma-Paulette PP, Bauer D (1997) Synthesis and properties of cerium and titanium oxide thin coatings for corrosion protection of 304 stainless steel. J Sol-Gel Sci Technol 10:317–331

NdFeB-info.com, 2015

Optiglass Limited (2015) Starna®, certified reference materials for UV and visible spectroscopy. Accessed Mar 2015

Patent US837017. https://www.google.com/patents/US837017

Preisler EJ (2003) Investigation of novel semiconductor heterostructure systems—I: Cerium oxide/silicon heterostructures—II: 6.1 Å semiconductor-based avalanche photodiodes. PhD thesis, California Institute of Technology, 168 pp

Rokhlin LL (2003) Magnesium alloys containing rare earth metals. Advances in metallic alloys, vol 3. CRC Press, Cleveland, 256 pp

Royal Society of Chemistry (2011) http://www.rsc.org/Education/EiC/issues/2011June/TheElements.asp. Accessed Mar 2015

Royal Society of Chemistry (2015) http://www.rsc.org/periodic-table/element/57/lanthanum. Accessed Mar 2015

Scandium.org (2015) http://www.scandium.org/Sc-Al.html. Accessed Mar 2015

Sherry AD, Caravan P, Lekinsky RE (2009) Primer on gadolinium chemistry. J Magn Reson Imaging 30:1240–1248

Smith GP (1967) Photochromic glasses: properties and applications. J Mat Sci 2(2):139–152

Strnat K, Hoffer G, Olson J, Ostertag W, Becker JJ (1967) A family of new cobalt-based permanent magnet materials. J Appl Phys 38(3):1001–1002

Vinogradova NN, Dmitruk LN, Petrova OB (2004) Glass transition and crystallization of glasses based on rare-earth borates. Glass Phys Chem 30(1):1–5

Wickersheim KA, Lefever RA (1964) Luminescent behavior of the rare earths in yttrium oxide and related compounds. J Electrochem Soc 111:47–51

Wikipedia (2015a) http://en.wikipedia.org/wiki/Carl_Auer_von_Welsbach

Wikipedia (2015b) http://en.wikipedia.org/wiki/Mischmetal

Wikipedia (2015c) http://en.wikipedia.org/wiki/Cerium(IV)_oxide

Wu MK, Ashburn JR, Torng CJ, Hor PH, Meng RL, Gao L, Huang ZJ, Wang YQ, Chu CW (1987) Superconductivity at 93 K in a new mixed-phase Y-Ba-Cu-O compound system at ambient pressure. Phys Rev Lett 58(9):908–910

Yung Y, Bruno K (2012) Low rare earth catalysts for FCC Operations. Pet Technol Q 1–10. http://www.digitalrefining.com/article/1000347

Yuriditsky B (2003) The crystallization mechanism of cerium-opacified enamels. In: Faust WD (ed) 65th Porcelain Enamel Institute Technical Forum: ceramic engineering and science proceedings, vol 24. Wiley, Hoboken, p 5. doi:10.1002/9780470294840.ch16

Chapter 6
Economic Aspects of the Rare Earths

Abstract In this chapter, the global production of the rare earths is discussed, expected shortages and surpluses for certain rare earths (the "balance problem") are indicated, and it is explained how China came to be the largest producer of rare earth elements in the world. Of course, the so-called Rare Earth Crisis of 2009–2013 is also addressed.

6.1 Introduction

The rare earths are used world-wide, and the availability and price of these metals has therefore world-wide effects. It happened that, from approximately the year 2000, production came largely from China. In 2009 the prices of the rare earths rose enormously (the so-called "Rare Earth Crisis"). This made the production by other countries worthwhile, or, in the case of the American Mountain Pass mine, made production again viable. This chapter explains more about these matters.

6.2 Global REE Production

The global production of the rare earth elements prior to 2010 is summarized in Fig. 6.1, and was approximately 110,000 tons per year in 2010.

Before 1960, the production of the rare earth elements was approximately 2 ktons per year. Production was mainly from monazite (and xenotime) from placer deposits (Geschneider 2011). The start of the growth of the rare earth industry began in the early 1960s, when it was discovered that the element europium (Eu) gave an intense red luminescence when exited by electrons. This was very quickly utilized in the development of color TV's (Geschneider 2011).

Looking at Fig. 6.1, it appears that industrial demand is in fact low in terms of ton. In 2012, the annual primary production was about two orders of magnitude less

J.H.L. Voncken, *The Rare Earth Elements*, SpringerBriefs in Earth Sciences,
DOI 10.1007/978-3-319-26809-5_6

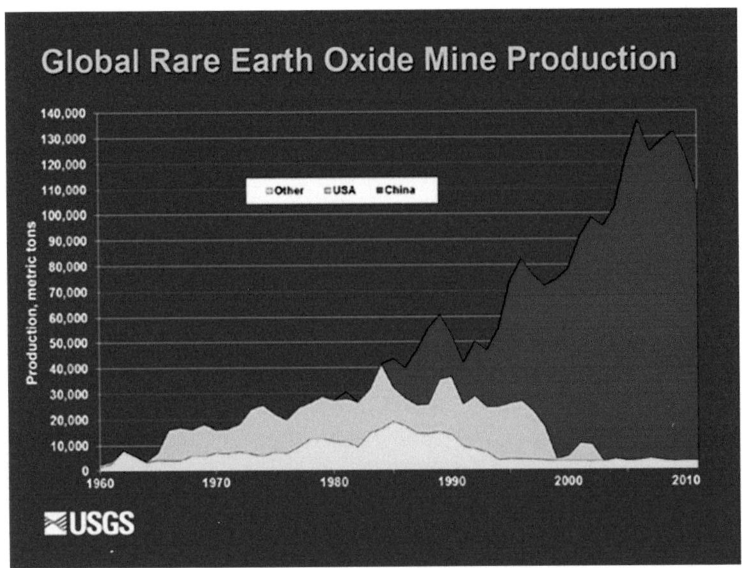

Fig. 6.1 Global rare-earth-oxide production trends. *Source* USGS (2015)

than copper and four orders of magnitude less than iron (Alonso et al. 2012). However, the commercial significance of the rare earth elements does not match the amount used. Several of the REEs are important because they are critical in a wide variety of applications. Also they are used in key technologies undergirding mobility and energy supply (Alonso et al. 2012).

The production of rare earth oxides from the Mountain Pass mine started in 1964 and remained the main source of light rare earths in the west until approximately the middle of the 1990s (Castor 2008). In approximately 1985, China began to export rare earth element concentrates, and by 1990, China was producing more than the USA (Geschneider 2011).

Since the middle of the 1990s, China has had a near monopoly in all aspects of the REE supply chain, including, production, processing, consumption, and R&D. Although small amounts of REE continued to be produced in Russia, India, and Brazil, by 2005. China supplied 97 % of the world's REE resources.

The global demand for REE in the year 2010 was approximately 110,000–130,000 tons per year (Fig. 6.1). The clear rise in demand for REE since 2000 is due to the value of REE in the fields of clean energy and high-technology products (Canadian Institute of Mining, Metallurgy and Petroleum 2015). Not all REEs are as high in demand as others. The highest commercial interest is for eight rare earth elements: lanthanum, cerium, neodymium, praseodymium, samarium, dysprosium, europium and terbium (Massari and Ruberti 2013).

Japan is the largest importer of REE, which amounts to about 73 % of the global demand outside of China. The European Union countries collectively import roughly 13 % and the United States imports roughly 3 % of China's REE production (Canadian Institute of Mining, Metallurgy and Petroleum 2015).

In the future, when the global supply/demand is projected out to 2020, a shortage of HREE (Tb, Dy, Er, Y) is expected. In 2016, the global demand for these HREEs is expected to be 14,500 tons, while production is expected to be 7000 tons (Canadian Institute of Mining, Metallurgy and Petroleum 2015).

Furthermore, the abundances of LREE and HREE are not similar. The LREEs are much more common. And within the lanthanide series, due to the Oddo-Harkins rule (see Chap. 3), the REEs with an even atomic number are more common than the REEs with an odd atomic number. In a chart of the abundance, this is clearly visible as a saw tooth pattern (Chap. 1, Fig. 1.3).

Then there is also the so-called "balance problem". The rare earths are found in nature as mixtures. The mixture found depends on the ore mineral, the type of ore, and the location of the deposit. Bastnaesite and monazite are minerals which are rich in the LREEs, whereas xenotime and the South-Chinese ion-adsorption clays are rich in the HREEs. Also, the REEs decrease in abundance with increasing atomic number Z. Now the "balance problem" is the balance between the abundance of the rare earth elements in ores and the demand on the economic markets (Binnemans et al. 2013).

A market in balance is very difficult to achieve because the demand for specific rare earths is different, due to different applications and technological innovations. This causes a sometimes very high demand for one REE, which is a minor element in the ore (for instance dysprosium), whereas the demand for another one, which is a major constituent in the ore (for instance yttrium) is much lower. To help solve this, one may tune the production of REEs in general to meet the high demand of a particular REE, and stockpile the other REEs with a lower demand. The latter leads to a price increase, due to the costs for stockpiling. Moreover, taking into account the operational margins of a producer, this leads to shortages of some REEs and surpluses of others (Binnemans et al. 2013).

It would be a preferable situation if the demand for elements that are very abundant would control the REE market. Unfortunately, this is not the case. The most wanted elements at this time are neodymium and dysprosium (Binnemans et al. 2013). Cerium, praseodymium, and the heavy REEs holmium, gadolinium, thulium, ytterbium and lutetium are produced in excess, and are stockpiled.

There are basically five solutions to the problem (Binnemans et al. 2013; Binnemans 2014).

1. Find new high-volume applications for the elements that are available in excess,
2. Find substitutions for those that are high in demand, but available in limited amounts.
3. Diversify the types of rare-earth ores mined, and use less common REE-minerals such as eudialyte or loparite. In the latter minerals, the concentration of REEs and the ratio of REEs occurring are different from what is found in the "common" ore minerals monazite, bastnaesite and xenotime.

4. REE-recycling. This is studied currently (2015) on a larger scale than before, due to the dependence on one major supplier (China). The main waste streams that are under consideration are REE permanent magnets, nickel-metal-hydride batteries, and lamp phosphors.
5. Reduced use. In a number of applications, smart engineering can make it possible to diminish the use of critical REEs, whereas performance is not reduced. A good example is NdFeB-magnets, which contain dysprosium to prevent demagnetization at higher temperatures. It has been established that dysprosium can diffuse to the grain boundaries of a sintered NdFeB-magnet. Due to this effect, while keeping a similar effect of dysprosium, less than 50 % of this metal is needed (Binnemans 2014).

6.3 How China Became the World's Largest REE Producer

In the 1980s, China started to develop innovative programs in science and technology. This resulted in two programs, which would accelerate the countries high-tech development. In March 1986, the leader of China at that time Deng Xiaoping approved Program 863: The National High Technology Research and Development Program. Program 863 focuses on biotechnology, space technology, information technology, laser technology, automation, energy technology, and on new materials. There are a mix of military and civilian projects in the program (Hurst 2010). Other programs are, for instance, the Nature Science Foundation of China (NSFC).

However, no programs have been so important as Program 863 and the later program 973. A very important researcher was Professor Xu Guangxian (1920–2015), who is called "*the father of rare earths in China*" (Peking University News 2015). China credits Xu with paving the way for the country to become the world's primary exporter of rare earth elements (Hurst 2010). Xu Guangxian applied his previous research in extracting isotopes of uranium to rare-earth extraction and succeeded in developing cutting-edge extraction technologies for the REEs.

Rare earths were the perfect materials to give China high profits and geopolitical influence (Bourzac 2011). Therefore, in the 1980s and 1990s, China decided it wanted to become a world leader in the production of rare earth elements. In the years 1978–1989, China increased production of REEs by an average of 40 % per year (Hurst 2010). In 1992, former vice-chairman of the CCP and vice-prime minister Deng Xiaoping said: "There is oil in the Middle East; there is rare earth in China." In 1999, President Jiang Zemin stated: "Improve the development and application of rare earths, and change the resource advantage into economic superiority" (Canadian Chamber of Commerce 2012). As China is a state lead economy, such a statement of Jian Zemin was made into reality.

While in the USA environmental regulations were very strict and labor costs relatively high, Chinese companies profited from a combination of low labor costs and lax environmental regulations. Also, the largest rare earth mine in China at Bayan Obo not only produces rare earths but also iron ore, which provides another stream of income which covers the mine's fixed costs (Gholz 2014).

In the 1990s, China's export of the rare earth elements grew, causing a significant world-wide drop in prices. As a result, other producers, such as Molycorp, became increasingly unprofitable (Hurst 2010). In 2002 the Mountain Pass mine shut down due to complaints about environmental damage. Also, the mine and associated processing plants needed capital investment, so there had been a laborious round of permit applications (Gholz 2014).

In 1990, China also started to export separated REE commodities. At the end of the 1990s, China produced not only separated REE-oxides and metals, but also produced higher value products, such as magnets, phosphors, and polishing compounds. Since approximately the year 2000, China also produces finished products, such as electric motors, computers, batteries, liquid crystal displays (LCD), mobile phones and portable music devices (Geschneider 2011).

The result of all this has been that China became one of the largest producers globally (Hurst 2010; Canadian Chamber of Commerce 2012).

Currently in China, there are two state key laboratories, which focus on rare earths. These are the State Key Laboratory of Rare Earth Materials Chemistry and Applications, affiliated with Peking University, and the State Key Laboratory of Rare Earth Resource Utilisation, affiliated with Chanchun University, and belonging to the Chinese Academy of Sciences (Hurst 2010). In 1987, the Open Laboratory of Rare Earth Chemistry and Physics was established, at the Changchun Institute of Applied Chemistry. In 2002, the laboratory was renamed, and in 2007 it became the State Key Laboratory of Rare Earth Resource Utilization (Hurst 2010).

This laboratory focusses on 3 main research fields:

- Solid state chemistry and the physics of the rare earths
- Bioinorganic chemistry and the chemical biology of rare earth and related elements
- Rare-earth separation chemistry

The laboratories and institutes each focus on a particular research area, but their research is complementary. With respect to the scientific literature, there are two journals that focus specifically on rare earths, and which originate in China: the Journal of Rare Earths (editor-in-chief was Xu Guangxian, Publisher: Elsevier), and the China Rare Earth Information Journal. These are the only two journals world-wide that focus almost exclusively on rare earths, and they are both run by scientists based in China.

6.4 The REE-Crisis (2009–2013)

China limited the export of rare earths in 2007, as they wanted to retain them for their own market. This was achieved by raising export duties. Export duties were originally at 10 %, but rose from 15 to 25 % in 2011. In 2011, China subjected the export of ferro-alloys, containing more than 10 % of rare earth elements, to taxes of 25 % (Stewart et al. 2011). Overall this resulted in a large drop in China's export of rare earths.

The effect of this was a strong rise of the prices of REEs on the world market (Fig. 6.2). This has been termed the REE Crisis.

In fact, the export taxes were a direct violation of China's WTO commitments (Stewart et al. 2011). In 2012, the US filed a protest against China's export taxes through the WTO. Soon other industrialized countries joined. The reason for this was that the US and other industrialized countries in the world wanted to produce rare-earth-containing devices for themselves, which required access to refined rare earths (CNN 2012).

In 2014, the WTO rejected China's main argument (the alleged reason for the reduction in exports was to conserve a limited resource and to reduce mining pollution). The WTO ordered China to remove the ceiling on exports of rare earths (Ferris 2015). China was ordered to cancel its export taxes on rare earth elements on 2 May 2015, in response to the 2014-decision of the WTO on the export taxes (Argus Rare Earths 2015).

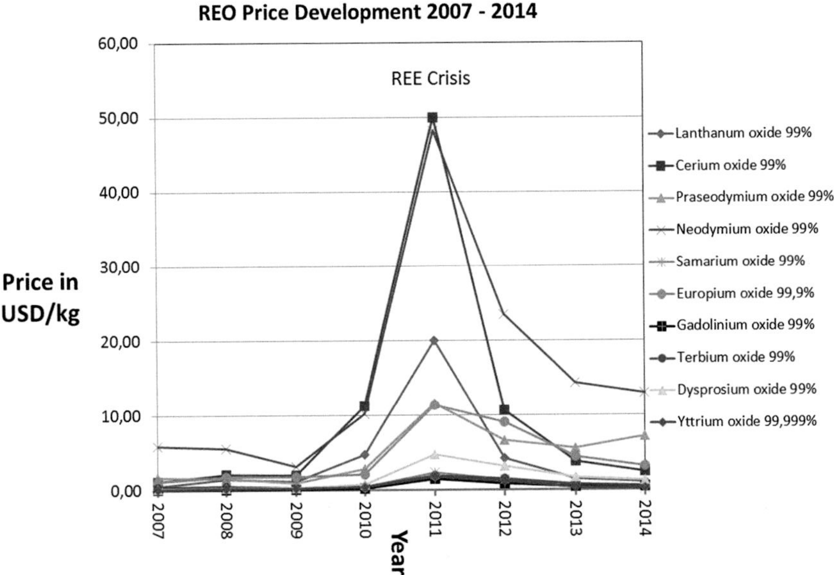

Fig. 6.2 REO price development 2007–2014. Based on data published by Metals Pages, and Lynas Corporation. NB: these are prices for REE-oxides (REO). Prices for the pure metals were much higher

China announced on April 24, 2015 the lifting of the export taxes (BBC 2015). Indeed China has done so. "Under the new guidelines, rare earth minerals will still require an export license in China but the amount that can be sold abroad will no longer be covered by quota" (Law360 2015).

The REE-crisis actually ended, however, before this date (see Fig. 6.2). This is because other producers emerged (e.g. Lynas Corporation) since the start of the REE-crisis, and Molycorp restarted its production. However, these resources mainly comprise the LREEs, with limited HREE. Non-Chinese mines with significant amounts of HREEs are expected to come into production after 2015–2016 (Canadian Institute of Mining, Metallurgy and Petroleum 2015). Examples of the latter are Norra Kärr, Sweden, which is especially rich in the HREEs, and which makes up more than 50 % of the total REE content of the deposit (Tasman Metals 2014), and Kringlerne, SW-Greenland (Tanbreez 2014). In the meantime, China will be the sole source of HREEs (Canadian Institute of Mining, Metallurgy and Petroleum 2015). It must be realized, that the lowering in the REO-prices has had severe consequences for producing companies outside China. Lynas corporation faces bankruptcy, and Molycorp has again stopped production at Mountain Pass, and has filed for bankruptcy (Mining.com 2015; The Sydney Morning Herald 2015).

References

Alonso E, Sherman AM, Wallington TJ, Everson MP, Field FR, Roth R, Kirchain RE (2012) Evaluating rare earth element availability: a case with revolutionary demand from clean technologies. Environ Sci Technol 46:3406–3414

Argus Rare Earths (2015) 17–22 Jan 2015, p. 4

BBC (2015) China scraps quotas on rare earths after WTO complaint. http://www.bbc.com/news/business-30678227. Accessed Aug 2015

Binnemans K, Jones PT, Van Acker K, Blanpain B, Mishra B, Apelian D (2013) Rare-earth economics: the balance problem. JOM 65(7):846–848

Binnemans K (2014) Economics of the rare earths: the balance problem. In: Proceedings of ERES2014: 1st European rare earth resources conference, Milos, Greece, pp 37–46. 04–07 Sept 2014

Bourzac K (2011) The rare-earth crisis. MIT Technol Rev Mag May/June 2011

Canadian Chamber of Commerce (2012) Economic policy series, pp 5–6. Apr 2012

Canadian Institute of Mining, Metallurgy and Petroleum (2015) http://www.cim.org/en/RareEarth/Home/GlobalReeProduction.aspx. Accessed May 2015

Castor SB (2008) The mountain pass rare earth carbonatite and associated ultrapotassic rocks, California. Can Mineral 46:779–806

CNN (2012) Obama announces WTO case against China over rare earths. http://edition.cnn.com/2012/03/13/world/asia/china-rare-earths-case/. Accessed 12 Mar 2012

Ferris D (2015) 5 years after crisis, U.S. remains dependent on China's rare earth elements. http://www.eenews.net/stories/1060011478. Accessed May 2015

Geschneider KA (2011) The rare earth crisis—the supply/demand situation for 2010–2015. Mater Matters 6(2):32–41

Gholz E (2014) Rare earth elements and national security council on foreign relations, 19 pp

Hurst C (2010) China's rare earth elements industry: what can the west learn? Institute for the Analysis of Global Security (IAGS), Washington DC, 42 pp

Law360.com (2015) China poised to scrap rare earth export duties. http://www.law360.com/articles/647264/china-poised-to-scrap-rare-earth-export-duties. Accessed Aug 2015

Massari S, Ruberti M (2013) Rare earth elements as critical raw materials: focus on international markets and future strategies. Resour Policy 38:36–43

Metals Pages (2015) http://www.metal-pages.com/. Accessed Apr 2015

Mining.com (2015) Molycorp shuts down mountain pass rare earth plant. http://www.mining.com/molycorp-shuts-down-mountain-pass-rare-earth-plant/. Accessed Aug 2015

Peking University News (2015) In memory of master Xu Guangxian. http://english.pku.edu.cn/news_events/news/people/3675.htm. Accessed May 2015

Stewart TP, Drake EJ, Dwyer AS, Gong P (2011) Rare earths, an update: a fresh look at the supplier(s), the buyers, and the trade rules: the global business dialogue, 35 pp. http://www.gbdinc.org

Tanbreez (2014) http://tanbreez.com/en/project-overview/tanbreez-%e2%80%93-what-is-it. Accessed Oct 2014

Tasman Metals Ltd. (2014) http://www.tasmanmetals.com/s/Norra-Karr.asp. Accessed Nov 2014

The Sydney Morning Herald (2015) Molycorp, sole US rare earth producer, files for bankruptcy. http://www.smh.com.au/business/mining-and-resources/molycorp-sole-us-rare-earth-producer-files-for-bankruptcy-20150627-ghzmsn.html. Accessed June 2015

USGS (2015) Rare earth productions trends. http://minerals.usgs.gov/minerals/pubs/commodity/rare_earths/. Accessed Mar 2015

Chapter 7
Recycling of Rare Earths

Abstract In this chapter, the need for recycling rare earths is discussed. It identifies the particular sources of materials that can be used for recycling, such as permanent magnets, lamp phosphors, CRT screens and flat-panel screens, polishing media, batteries, and bulk waste products such as red mud, a waste product from the Bayer process, and phosphogypsum, a by-product from the phosphoric acid production. As recycling of rare earths is not yet done to a large extent, some pro's and con's of recycling of rare earths are also discussed: why would you do it or not do it?

7.1 Introduction

Rare earth elements have an essential role in permanent magnets, lamp phosphors, rechargeable NiMH batteries, catalysts and other applications (see Chap. 5). In 2010, the European Union published the report "Critical Raw Materials for the EU", indicating that the rare earths had the highest supply risk of all raw materials, while their economic importance was considered medium (European Union 2010) (Fig. 7.1).

The US Department of Energy (DOE) indicated in 2011 that five of the rare earths are most critical for clean-energy purposes (Fig. 7.2.): neodymium (Nd), europium (Eu), terbium (Tb), dysprosium (Dy), and yttrium (Y), (U.S. Department of Energy 2011). They indicate some differences for the short-term future (at that time 2011–2015) and the medium-term future (at that time 2015–2025).

Even though China has lifted the taxes on REE export (see Chap. 6), its export especially of the heavy REEs is critical to western economies, since China has a near monopoly on the production of these metals.

Alonso et al. (2012) made predictive models for a broad range of scenarios of future REE demand and tried to draw conclusions from observations that come from the results of these scenarios. For Nd and Dy, they conclude that over the next 25 years the demand will rise by 700 and 2600% respectively.

J.H.L. Voncken, *The Rare Earth Elements*, SpringerBriefs in Earth Sciences,
DOI 10.1007/978-3-319-26809-5_7

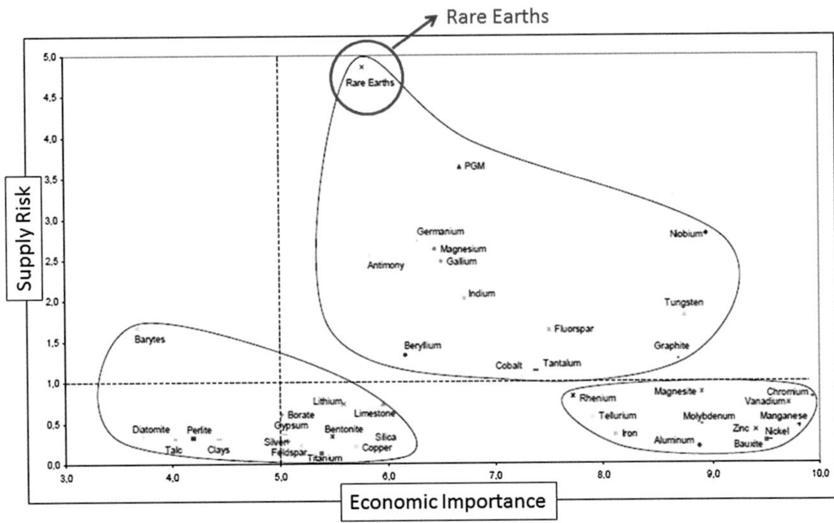

Fig. 7.1 Supply risk versus economic importance. The rare earths are in the *red circle*. Modified after the European Union (2010). Used with permission

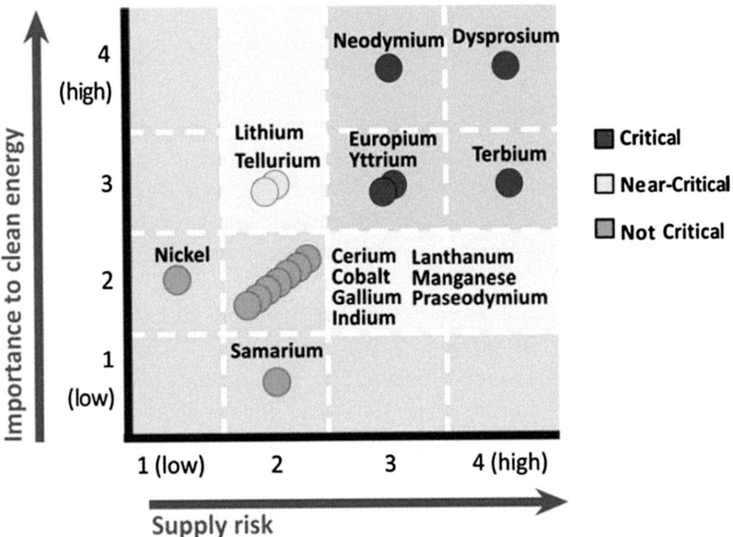

Fig. 7.2 DOE medium-term (2015–2025) criticality matrix, showing the five most critical rare earth elements (Y, Nd, Eu, Tb, Dy), and the non-critical ones (Ce, Pr, Sm). Reproduced with permission of the U.S. Department of Energy (2011)

Kifle et al. (2013) conclude that for rare earth elements, over the long-term (30–300 years), there will be an issue with long-term sustainability of supply, which can potentially be prevented by increasing recycling from waste.

7.2 Sources for Recycling

What sources for recycling are there? Considering critical-metal-containing waste streams, one may consider (Binnemans et al. 2015):

(1) Direct recycling of metal scrap and swarf which is formed during the production of metal-based final or intermediate products (for instance NdFeB-magnets).
(2) Postconsumer materials (recycling and urban mining): complex, multi-material, metal-containing products (for example hybrid electric vehicles).
(3) Landfill mining of historic urban solid waste.
(4) Metal recovery from industrial-process-flow residues from primary and secondary metal production.
(5) Metal recovery from stocks of landfilled industrial-process residues.

At present, most attention is focussed on the direct recycling of scrap and the urban mining and subsequent recycling of end-of-life REE-containing products (Binnemans et al. 2015).

The most interesting end-of-life products for the recycling of (heavy) rare earths are permanent magnets, lamp phosphors, CRT and flat-screen phosphors, and rechargeable batteries (Ni-metal hydride batteries). Much less attention has been paid to stocks which in the past were landfilled and flows of industrial process residues. These sources are relatively poor for rare earths, but constitute enormous volumes of material, so that the total amount of rare earths is also large. They may thus constitute an independent source of rare earths (Binnemans et al. 2013). Poscher et al. (2014) describe processes for recycling of rare earths from spent polishing powders. Recycling of these powders is a complicated and expensive process, and is probably only worthwhile when prices of the rare earths used in these materials (mainly lanthanum and cerium) are relatively high (Posher et al. 2014). Finally there are large volume waste products like "red mud", a residue from aluminium production, and phophogypsum, a residue from phosphorus production.

7.2.1 Permanent Magnets

Of the permanent magnets, recycling is carried out mainly for NdFeB-magnets, as these constitute most of the permanent REE-containing magnets. SmCo-magnets have presently a market share of less than 2 % (Binnemans et al. 2013). The life cycle of NdFeB-magnets depends on the application and varies from as short as 2–3 years in consumer electronics to as long as 20–30 years in wind turbines (Yang 2014). Therefore, most of the presently redundant NdFeB material is found in electronic apparatuses such as loudspeakers, mobile phones and hard disk drives (HDD) (Binnemans et al. 2013; Yang 2014). Over 80 % of the REE-content in NdFeB is in the form of neodymium and dysprosium (Du and Graedel 2011).

Of these, the largest mass flow is constituted by HDDs. More than 200 companies have manufactured HDDs over time. But over the years, the production has today become concentrated into just three manufacturers: Western Digital, Seagate, and Toshiba. The amount of HDDs shipped annually since 1976 is shown in Fig. 7.3, with currently some 600 million manufactured annually (Walton and Williams 2011; Binnemans et al. 2013).

Recycling of NdFeB-magnets from post-consumer HDDs is described in detail by Abrahami et al. (2015). They learned that, for crushed HDDs, the REEs were

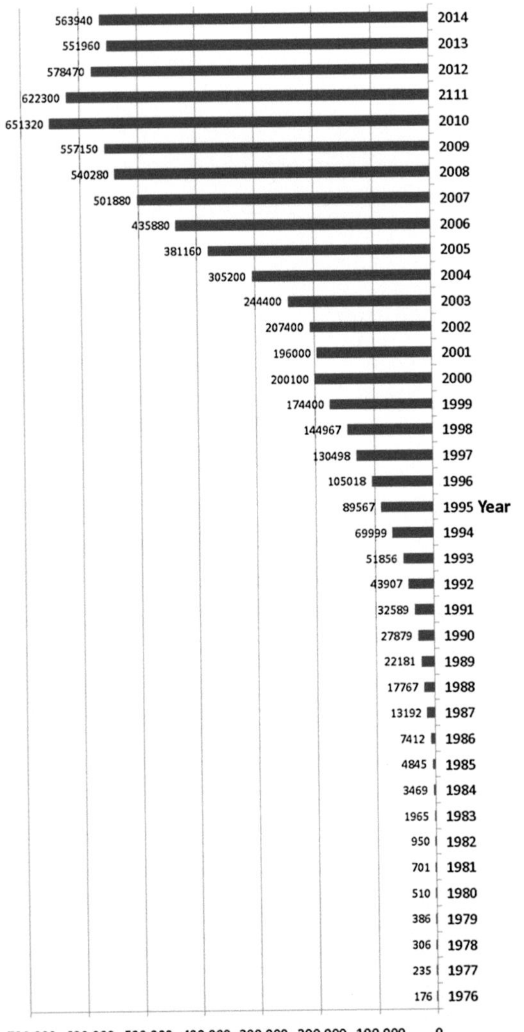

Fig. 7.3 Amounts of HDD shipped world-wide 1976–2014 (*Data source* Storage Newsletter 2015)

found in all size dimensions below 1 mm, which was 63 % of the entire grate residue. A mixture of all these size fractions was made, and labelled upgraded scrap. They describe a method where scrap was leached with 2M H_2SO_4, similar to the method described by Lyman and Palmer (1993) for recycling NdFeB-magnet scrap from magnet manufacture. They were able to remove 98 % of the Nd from the scrap. A pyrometallurgical method using molten salt slags was less successful (Abrahami et al. 2015).

7.2.2 Lamp Phosphors

Luminescent materials, also called phosphors, are mostly solid inorganic materials consisting of a host lattice, usually intentionally doped with impurities. The absorption of energy takes place via either the host lattice or the impurities. In almost all cases, the emission originates from impurities (Ronda et al. 1998).

Luminescent materials are widely applied today (e.g., in emissive displays and fluorescent lamps). In Fig. 7.4, some fluorescent lamps are shown. Although the major recycling incentive in the past was the safe removal of the mercury present in the lamps (Silveira and Chang 2011), recent developments are also focussed on recycling the REEs (Dupont and Binnemans 2015). Rare earth elements used in fluorescent lamps are La, Ce, Eu, Tb and Y (Philips Lighting Company 2011;

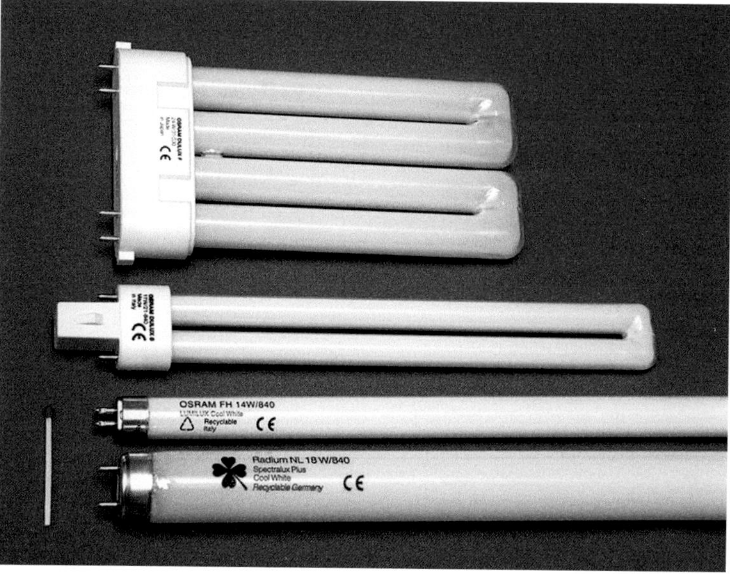

Fig. 7.4 Fluorescent lamps (and a match for comparison). *Source* Wikimedia commons. Original photograph by Christian Taube

Fluorescent Lamp Stewardship Initiative 2000; European Lamp Companies Federation 2015; Binnemans and Jones 2014). Of these, Eu, Tb and Y are critical REEs (see Fig. 7.2).

Wu et al. (2014) reviewed several processes to recover rare earth elements from waste fluorescent lamps. The reviewed processes involve wet crushing using a solution containing 30 % acetone (Rabah 2008) and dry crushing. Dry crushing is carried out in a closed environment or vacuum environment, because of the hazardous mercury vapors. After the crushing step, dissolution follows, which is carried out either by acid leaching using sulphuric acid (leaching with HCl and HNO_3 gave a lower efficiency) or by alkali fusion. Alkali fusion is the thermal decomposition of insoluble substances. With this method is aimed at the destruction of the structure of substances and achieving the transition to soluble substances. In the next step, the transfer of the rare-earth ions from a soluble state into insoluble precipitation is conducted by adding oxalic acid or oxalate. After this follows the extraction step, where solvent extraction[1] and supercritical fluid extraction[2] (SFE) are usually used (Wu et al. 2014). Today about 680 million lamps are disposed of annually in the USA alone, mostly to solid-waste-disposal facilities, including landfills and solid-waste incinerators (State of Washington 2015).

7.2.3 CRT Screens and Flat-Panel Screens

Although CRT displays, used in devices such as computer monitors and TV tubes, have been replaced by new technologies such as LCD displays, plasma, LED and other flat screen technologies, there is a growing amount of electronic waste containing these. In CRT screens, REEs are used in phosphors (substances that emit luminescence). Yttrium, europium, and terbium phosphors are the red-green-blue phosphors used in many light bulbs, panels, and televisions (USGS 2014). Yttrium and europium, for instance, are essential for the color in flat computer screens, cell phones, and vehicle dashboards (IPC 2015).

Resende and Morais (2015) present a study of the coating-powder leaching from spent CRT-computer-monitor screens and TV-tubes to recycle europium and

---

[1]Solvent extraction or liquid-liquid solvent extraction uses two immiscible or partially immiscible solvents containing dissolved rare earths. The two liquids are mixed, the solutes are allowed to distribute between the two phases until equilibrium is established, and then the two liquids are separated. The concentrations of the solutes in the two phases depend upon the relative affinities for the two solvents. According to convention, the product (liquid) that contains the desired solute is called the "extract," while the residue left behind in the other phase is called the "raffinate." (Encyclopedia Britannica 2015a).

[2]Supercritical fluid extraction (SFE) is the process for large-scale purification of complex liquid or solid matrices, separating one component (the extractant) from another (the matrix) using supercritical fluids as the extracting solvent. The supercritical fluid is usually CO_2. The major advantage of this method over liquid-liquid extraction is that the supercritical fluid can easily be removed after extraction by lowering the temperature or pressure or both. (Encyclopedia Britannica 2015b).

yttrium. First, the material is digested by sulphuric acid, followed by dynamic water leaching[3] at room temperature. In the CRTs, yttrium is present as Y_2O_2S (yttrium oxysulphide), and europium is a dopant. $Y_2O_2S{:}Eu^{3+}$. This is the red phosphor component. As the acid digestion proceeds, a trivalent europium and yttrium sulphate is formed, while H_2S is liberated. In a second step, by dynamic leaching using water, metals are leached from the solid produced in the acid digestion step.

Flat-panel screens contain mercury (except for plasma screens). These flat panel screens require separate internal lighting sources to illuminate the liquid-crystal flat screens from behind. Usually this is a cold-compact fluorescent lamp (CCFL) which contains mercury.

They are used in appliances including television sets, computer monitors, mobile phones, handheld video game systems, personal digital assistants, and navigation systems (e.g., Tasman Metals 2015) see also Chap. 5.

Recycling of these types of monitors is not (yet) common, at least not for rare earth elements, as rare earth elements are used in very tiny amounts. For instance, Ayres et al. (2014) mention that the amount of europium in LCDs using CCFLs varies from 8.10 mg for a TV to 0.13 mg for notebook computers, whereas LCDs using LEDs contain an even lower amount: 0.09 mg for TVs compared to 0.03 mg for notebooks.

7.2.4 Polishing Media

The polishing materials and additives used in the glass industry are estimated to be 10 % of the global demand for rare earths in terms of economic value. However, these applications require mainly cerium and lanthanum (Posher et al. 2014). Cerium and lanthanum are the most abundant REEs and the continuously falling prices (see Chap. 6, Fig. 6.2) in the last years makes economic recycling difficult to justify. Recycling processes therefore should be easy manageable, environmentally friendly and low-cost. On the other hand, the strategic independence of the European Union and many other countries from China's supply of rare earth elements is a driving force for recycling research (Posher et al. 2014).

Several investigators have proposed methods to recycle glass polishing powder. Kato et al. (2000) used a 4 mol/kg NaOH solution and temperatures of 50–60 °C on the spent-glass polishing powder. Major impurities in the waste are a SiO_2 component resulting from fine glass powder and Al_2O_3 component resulting from coagulating agents. These impurities react with NaOH to precipitate as zeolite at higher temperatures. Other methods were patented by Moon et al. (2011) and by Matsui et al. (2013).

Moon et al. (2011) describe a method for recycling a cerium oxide abrasive by making a slurry of the cerium oxide. Subsequently there is a step including addition

[3]In dynamic leaching, the leaching solution is continuously renewed.

of a strong alkali solution (NaOH or KOH), a step of adding sodium fluoride, and a step of separating particles. The addition of the alkali solution leads to the removal of a Si-OH layer and also a fine powder of glass material from the Ce-oxide particles. Then there is addition of sodium fluoride to the slurry waste, which gives a better separation of the component treated. The Ce-oxide particles may then be separated using, for instance, sedimentation, centrifugal separation, specific gravity separation, flotation, or filtration.

Matsui et al. (2013) propose a method that is partly similar, but they use as precipitants aluminium sulphate and polyaluminum chloride.

7.2.5 Nickel-Metal-Hydride Batteries

In the past, a very popular type of rechargeable battery contained cadmium (Cd), which is a very toxic metal. Cadmium is now banned by the EU for most uses. In 2006, the official Journal of the European Union in Directive 2006/66/EC, published Directive 2006/66/EC, article 4, stipulating that portable batteries or accumulators, containing more than 0.002 % of cadmium, are prohibited (European Union 2006). Although there are several types of rechargeable batteries, such as lithium-ion (Li-ion) and lithium-ion-polymer (Li-ion polymer), nickel-metal-hydride batteries (NiMH-batteries) are a very common type.

NiMH batteries have applications in common household applications (radios, remote controls, digital cameras, portable DVD players etc.), but are also used in electric and hybrid cars (Figs. 7.5 and 7.6).

A process for recycling such batteries was invented jointly by Umicore and Rhodia. This process combines Umicore's patented Ultra-High-Temperature battery recycling process (Umicore 2015) with the REE-refining capabilities of Rhodia. This process can be used for the whole range of NiMH batteries, from

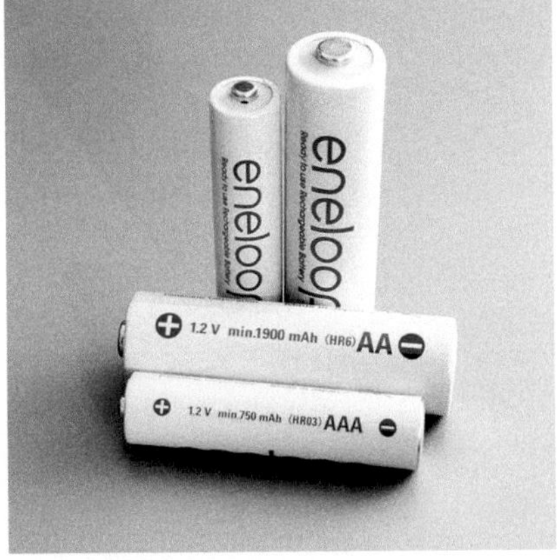

Fig. 7.5 Household NiMH batteries. *Source* Wikimedia commons. Original photograph by Ashley Pomeroy

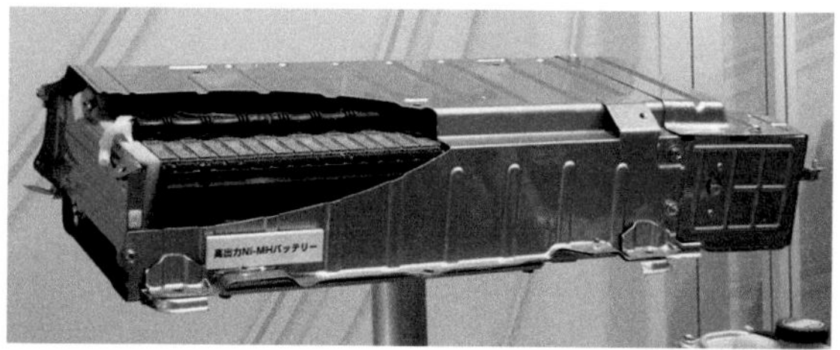

Fig. 7.6 Rechargeable NiMH battery for a Toyota Prius. *Source* Wikimedia commons. Original Photograph by Hatsukari715

batteries for portable applications to batteries for hybrid electric cars. Large batteries, like those from hybrid and fully electric cars, are first dismantled in a dedicated dismantling line. Umicore operates one dismantling line in Hanau (Germany) and another one Maxton, NC (USA). The smelting operations are performed in Hoboken (Belgium). In this process, the REEs contained in the NiMH batteries are collected in the slag fraction. They can be further refined to produce pure REE oxides. This is carried out by French company Rhodia, which in turn is owned by French chemical giant Solvay. Rhodia is among other things specialized in refining rare earths (Rhodia 2015).

Tang et al. (2013) studied the recycling of NiMH household batteries. The main parts of a NiMH battery are the cathode, the anode, the electrolyte, a separator, and the steel case. REEs are found in the anode, which consists of a hydrogen storage alloy based on misch metal and nickel alloys. The misch metal contains mainly cerium, lanthanum, praseodymium and neodymium.

The household batteries are recycled as follows (Tang et al. 2013). The batteries are put into liquid nitrogen for about 15 min. After that, they are immediately crushed in a jaw crusher. The crushed batteries are further treated by magnetic separation. Two separation steps are used to separate the steel scraps and the Ni-based alloys. Next, the plastic sealing plate and polymer separator have to be separated from the remaining parts of the positive and negative electrodes. For this, the material is heat treated in a muffle furnace. After trying several temperatures, 600 °C was chosen, to remove the plastic and polymer as efficiently as possible, without oxidizing the Ni-alloy too much.

After heat treatment, a sieving procedure is applied to separate the black powder containing the REEs from the rest of the material. XRF-analysis of the black powder shows it to contain about 22.1 % of REE oxides (mainly cerium, lanthanum and neodymium oxides). NiO content is 53.6 % and CoO content is 7.7 %.

After this, the materials are further separated by a slagging process. A calcium silicate slag is used in this process, which is conducted in a vacuum furnace at 1700 °C. This slagging process converts the material into a NiCo-alloy and a REE-oxide rich slag, which can be further processed for extraction of the REEs (Tang et al. 2013).

7.2.6 *Bulk-Waste Products*

There are two bulk-waste products from which REEs can be recycled. They are "red mud", a bauxite residue generated during the Bayer process for alumina production and phosphogypsum, a residue that is formed during phosphorus production, and contains REE originating from the original phosphorus ore mineral (apatite).

Recycling of REEs from those products has also been investigated. Borra et al. (2015) used organic and mineral acids (HCl, HNO_3, H_2SO_4, CH_3COOH, CH_3SO_3H and citric acid). They studied the effect of acid type and concentration, leaching time, liquid-to-solid ratio, and temperature. The concentration of the rare-earth elements (REEs) in the studied bauxite residue was around 0.1 wt%. Extraction of the REEs was the highest for HCl leaching compared to other acids, but the iron leaching was also high (about 60 %).

Koopman and Witkamp (2000) investigated the leaching of REE from phosphogypsum using sulfonic-acid ion-exchange resin. Although not primarily focussed on the lanthanides, their process is able to remove up to 53 % of the lanthanide input.

7.3 Recycling, Pros and Cons

Recycling of rare earths is until recently rarely undertaken in practice (Binnemans and Jones 2014; Graedel et al. 2011). Marshall (2014) points out several reasons why. These elements are present in only very small amounts in things like cell phones, and, as parts get smaller, so do the amounts of material used. In a touch screen, the elements are distributed throughout the material at the molecular level, and thus are not present as separate parts.

Recycling rare earth magnets (NdFeB) recovers the material mostly in the composition it was originally in, e.g., $Nd_2Fe_{14}B$, while REEs used as glass colorant require expensive processing to produce them in the oxide form (in which they are usually traded). This, however, does not take into account the amounts of minor additions to the magnet material for specialist use, for instance dysprosium and terbium.

To recycle REEs from scrap often means using aggressive solvents, and high temperatures, or even high pressures (Gupta and Krishnamurthy 2005). Because of the harmful materials and the large amounts of energy, recycling could mean greater environmental harm than mining virgin material (Marshall 2014).

On top of that, apart from the challenge of the recycling processes, there is the challenge of collecting the materials. And as the prices of the rare earths have dropped dramatically since 2011 (see Chap. 6), the pressure to recycle the REEs has dwindled, and only government regulation for recycling drives continuation of this in countries where strict recycling legislation is in force.

Often recovery of REEs, where undertaken, is not the primary goal of recycling. For instance, fluorescent light bulbs are recycled in the first place for the mercury in the bulbs, and electronics are recycled for the precious metals used in the equipment, such as gold, palladium and iridium (Marshall 2014). Therefore, recycling of REEs is probably more a matter of strategic-resource policy than of actual benefit in a straightforward financial or ecological way. For instance, the Raw Materials Initiative of the European Union (2008) states that the European Union is highly dependent on imports of metallic materials and that the EU relies heavily on secondary materials. The REEs are here mentioned especially, alongside metals such as cobalt, platinum and titanium.

References

Abrahami ST, Xiao Y, Yang Y (2015) Rare-earth elements recovery from post-consumer hard-disc drives. In: Mineral processing and extractive metallurgy. Trans Inst Min Metall C 124(2):106–115

Alonso E, Sherman AM, Wallington TJ, Everson MP, Field FR, Roth R, Kirchain RE (2012) Evaluating rare earth element availability: a case with revolutionary demand from clean technologies. Environ Sci Technol 46:3406–3414

Ayres RU, Villalba Mendez G, Talens Peiro L (2014) Recycling rare metals. Chapter 4 in handbook of recycling. Elsevier Inc., Ernst Worrell and Markus Reuter, eds.: 27–38

Binnemans K, Jones PT, Blanpain B, Van Gerven T, Yang Y, Walton A, Buchert M (2013) Recycling of rare earths: a critical review. J Clean Prod 51:1–22

Binnemans K, Jones PT (2014) Perspectives for the recovery of rare earths from end-of-life fluorescent lamps. J Rare Earths 32(3):195–200

Binnemans K, Jones PT, Blanpain B, Van Gerven T, Pontikes Y (2015) Towards zero-waste valorisation of rare-earth-containing industrial process residues: a critical review. J Clean Prod 99:17–38

Borra C, Pontikes Y, Binnemans K, Van Gerven T (2015) Leaching of rare earths from bauxite residue (red mud). Miner Eng 76:20–27

Du X, Graedel TE (2011) Global rare earth in-use stocks in NdFeB permanent magnets. J Ind Ecol 5(6):836–843

Dupont D, Binnemans K (2015) Rare-earth recycling using a functionalized ionic liquid for the selective dissolution and revalorization of Y_2O_3:Eu^{3+} from lamp phosphor waste. Green Chem 17(2):856–868

Encyclopedia Britannica (2015a) Solvent extraction. http://www.britannica.com/science/rare-earth-element/Ion-exchange#ref1177514. Accessed Aug 2015

Encyclopedia Britannica (2015b) http://www.britannica.com/science/separation-and-purification/Exclusion-and-clathration#ref619630. Accessed Aug 2015

European Lamp Companies Federation (2015) http://www.elcfed.org/2_health_environment.html#tech Accessed June 2015

European Union (2006) Directive 2006/66/EC of the European Parliament and of the Council of 6 Sept 2006 on batteries and accumulators and waste batteries and accumulators and repealing Directive 91/157/EEC. Official J Eur Union, L 266/5. http://eur-lex.europa.eu/LexUriServ/LexUriServ.do?uri=OJ:L:2006:266:0001:0014:EN:PDF

European Union (2010) Critical raw materials for the EU. Report of the Ad-hoc Working Group on defining critical raw materials. European Commission, Enterprise and Industry, Directorate General, 85 pp

Fluorescent Lamp Stewardship Initiative (2000) By Randall Conrad & Associates Ltd., 22 pp. http://environment.gov.ab.ca/info/library/6344.pdf

Graedel TE, Allwood J, Birat J-P, Buchert M, Hagelüken C, Reck BK, Sibley SF, Sonnemann G (2011) What dowe know about metal recycling rates? J Ind Ecol 15(3):355–366

Gupta CK, Krishnamurthy N (2005) Extractive metallurgy of the rare earths, CRC Press, 484 pp

IPC (2015) Future supply of rare earth elements—IPC—association connecting electronics industries. http://www.ipc.org/ContentPage.aspx?pageid=future-supply-of-rare-earth-elements. Accessed June 2015

Kato K, Yoshioka T, Okuwaki A (2000) Study for recycling of ceria-based glass polishing powder. Ind Eng Chem Res 39:943–947

Kifle D, Sverdrup H, Koca D, Wibetoe G (2013) A simple assessment of the global long term supply of the rare earth elements by using a system dynamics model. Environ Nat Resour Res 3 (1):77–91

Koopman C, Witkamp GJ (2000) Extraction of lanthanides from the phosphoric acid production process to gain a purified gypsum and a valuable lanthanide by-product. Hydrometallurgy 58:51–60

Lyman JW, Palmer GR (1993) Recycling of rare earths and iron from NdFeB magnet scrap. High Temp Mater Process (Lond) 11(1–4):75–88

Marshall J (2014) Why rare earth recycling is rare. http://ensia.com/features/why-rare-earth-recycling-is-rare-and-what-we-can-do-about-it/. Accessed July 2015

Matsui H, Harada D, Takeuchi M (2013) Method for recovery of cerium oxide, Patent US2013/0152483

Moon WJ, Na SO, Oh HY (2011) Method for recycling cerium oxide abrasive, Patent US2011/0219704

Philips Lighting Company (2011) Phosphor—a critical component in fluorescent lamps. Brochure, 11 pp

Posher A, Luidold S, Schnideritsch H, Antrekowitsch H (2014) Extraction of lanthanides from spent polishing agent. In: Proceedings of ERES2014—1st European rare earth resources conference, Milos, Greece, 04–07 Sept 2014, pp 457–466

Rabah MA (2008) Recyclables recovery of europium and yttrium metals and some salts from spent fluorescent lamps. Waste Manag 28:318–325

Resende LV, Morais CA (2015) Process development for the recovery of europium and yttrium from computer monitor screens. Miner Eng 70:217–221

Rhodia (2015) http://www.rhodia.com/en/markets_and_products/product_finder/index.tcm and http://www.rhodia.com/en/markets_and_products/product_finder/product_results.tcm?ProductRange=Rare+earths%2C+mixed+oxides+%26+aluminas%2FRare+earths

Ronda CR, Jüstel T, Nikol H (1998) Rare earth phosphors: fundamentals and applications. J Alloys Comp 275:669–676

Silveira GTR, Chang S-Y (2011) Fluorescent lamp recycling initiatives in the United States and a recycling proposal based on extended producer responsibility and product stewardship concepts. Waste Manage Res 29(6):656–668

State of Washington (2015) Mercury-containing lights and lamps as universal waste. Department of Ecology. http://www.ecy.wa.gov/programs/hwtr/dangermat/universal_waste_lamps.html Accessed Aug 2015

Storage Newsletter (2015) http://www.storagenewsletter.com/rubriques/market-reportsresearch/564-million-hdds-shipped-in-2014-trendfocus/. Accessed June 2015

Tang K, Ciftja A, van der Eijk C, Wilson S, Tranell G (2013) Recycling of the rare earth oxides from spent rechargeable batteries using waste metallurgical slags. J Min Metall Sect B 49B (2):233–236

Tasman Metals (2015) Principal uses of rare earth elements. http://www.tasmanmetals.com/s/PrincipalUses.asp Accessed Aug 2015

Umicore (2015) http://www.batteryrecycling.umicore.com/UBR/process/. Accessed Aug 2015

U.S. Department of Energy (2011) 2011 Critical Materials Strategy

USGS (2014) The rare-earth elements—vital to modern technologies and lifestyles. USGS Mineral Resources Program. Fact Sheet 2014–3078. ISSN 2327–6932 (online). http://dx.doi.org/10.3133/fs20143078

Walton A, Williams A (2011) Rare earth recovery. Mater World 19:24–26

Wu Y, Yin X, Zhang Q, Wang W, Mu X (2014) The recycling of rare earths from waste tricolour phosphors in fluorescent lamps: a review of processes and technologies. Resour Conserv Recycl 88:21–31

Yang Y (2014) Recovery of rare earth metals from end-of-life permanent magnets: an overview. In: Proceedings of ERES2014—1st European rare earth resources conference, Milos, Greece, 04–07 Sept 2014, pp 429–445